Essays on the Frontiers of Modern Astrophysics and Cosmology

Santhosh Mathew

Essays on the Frontiers of Modern Astrophysics and Cosmology

 Springer

Santhosh Mathew
Regis College
Weston, USA

SPRINGER–PRAXIS BOOKS IN POPULAR ASTRONOMY

ISBN 978-3-319-01886-7   ISBN 978-3-319-01887-4 (eBook)
DOI 10.1007/978-3-319-01887-4
Springer Cham Heidelberg New York Dordrecht London

Library of Congress Control Number: 2013949140

Cover illustration: Cover photo courtesy of Creative Commons

Printed on acid-free paper

Springer is part of Springer Science+Business Media (www.springer.com)

To the memory of Noah Mathew, who departed us 3 years ago.

Foreword 1
By Debra Leahy

What if our construct of time ran circular, or from side to side, rather than in a strictly linear fashion? What if a movement as small and precise as a butterfly flapping its wings in South America affected the weather in Central Park? These questions represent just a sampling of the provocative inquiries posed by Santhosh Mathews as he leads his readers through a journey as much about philosophy as it is about science, and, ultimately, about the human abilities of perception and imagination.

Central to Mathew's narrative is his weaving of Eastern philosophical beliefs, most notably Hinduism, and scientific reasoning. For some, the blending of these will be novel, while for others, this book will mark a continuation of a journey to understand how these two perspectives influence, counter, and complement one another. Throughout the book, scientific voices and views are brought to the forefront, such as Hawkins, Penrose, Newton, Dirac, and Einstein. While the invocation of these essential scientific thought leaders is not surprising, the seamless ways in which they are interwoven with a diverse array of voices and images, ranging from Shakespeare to Faulkner, from Pope to Frost, and from Picasso to Dali, along with the philosophical underpinnings associated with Brahma, Shiva, and Vishnu, are both enlightening and inspiring.

Science and philosophy are often perceived as opposite lines of thought; however, isn't it through the juxtaposition of seemingly opposing ideas that the true existence of both becomes more vivid? In bringing these voices together, Mathew shows how, in many cases, these modes of thought have been shown to be mirrors of one another. For instance, Mathew discusses how human, universal themes, such as the desire for good to prevail over evil, appear in science, through the analysis and beliefs related to matter versus anti-matter, and are also reflected in Hinduism, such as the conflict between Devas and Asuras, and additionally all the way up to contemporary fiction, such as *Angels and Demons*. By positioning these universal themes side by side, reflective images are revealed and we begin to view them as a continuum that narrates the depth and breadth of human history.

Mathew also demonstrates how, in other cases, philosophical beliefs and scientific reasoning have proven to be windows to one another. As patterns are detected and explored in science, and are checked against philosophy, scientific hypotheses gather further validity.

And, as scientific ideas and discoveries draw humans closer to universal truths, the infinite possibilities inherent to philosophy remind us that the answers that we seek are not arrived upon with any semblance of ease and may, in fact, be beyond our grasp. Perhaps the greatest homage that Mathew pays to the window offered through philosophy occurs when he reminds us that many scientists have turned to ancient Eastern texts to test their assumptions and widen and expand their imaginative sources of inquiry.

Readers are reminded of the small, insignificant place that humans hold in the vast expanse of the universe. Throughout the book, humans are the actors who advance science, but, at the same time, readers are continually reminded that the roles played by humans are minute and transient. How can we fully explain existence and meaning when we cannot even fully discern when life, as we know it, began? Although humans are imbued with a narrative that spans back far before their individual birth, the existence of any individual person will leave only an infinitesimal mark on a universe that is infinite.

Although critical scientific history and significant moments of discovery are clearly discussed throughout the book, Mathew continually juxtaposes the satisfaction of discovery and new knowledge with the wonder and beauty of the undiscovered and the unanswered; in doing so, he often holds the latter up as being more divine. Mathew's book presents its readers with a view of the universe as infinite, inviting, and inspiring. Yet, while the universe that Mathew presents is magnificent, it contains a deep narrative that is not yet fully written.

Mathew shows that humans have an inherent desire for knowledge and truth; yet, this has only led to more questions than answers. It is human nature to seek empirical truths and to crave outcomes. Yet, Mathew shows that *enlightened* humans desire questions over simple answers and see beauty and opportunity within continual inquiry. In this latter, more enriched quest, answers are found in less than expected places. Can we understand more from silence than sound, more from what is absent rather than what is present, and more from ancient texts than cutting-edge information? Mathew ponders all of these questions and invites the readers to do the same.

However, in Mathew's journey, there are no absolute outcomes or answers, and, in that space, we rejoice. Beauty lives in inquiry, and it lives in process over outcomes; it lives in wondering not how something works but why something works. Imagine a world where we are more comforted by what we don't know than what we do know. Imagine a world where skepticism was the positive choice that ignited imagination in infinite ways. Imagine a world where a simple algebraic equation holds answers to some of life's most pressing existential questions. Imagine a world where we do not accept convention with ease and where we are not held captive to time, measurement, distance, or borders. Mathew envisions that world and presents it to his readers throughout his book, and we are better for seeing it in such a fluid, unfixed way and through multiple lenses.

In a period where we are experiencing access to more information than ever before, this book is a gift to the reader who wishes to be a responsible steward of that immense array of information and use it to advance inquiry rather than accept the semblance of truth so readily at our fingertips. Mathew extends an invitation to take this mode of inquiry and pass it to the generations beyond us. Like his aptly titled chapter that begins with "Once upon a time," his narrative is meant to be passed on.

What will future generations think when they ponder the mark that we have left, albeit small in scope, on this universe? With each chapter we add or subtract a brick to our warehouse of collective understanding, Mathew suggests that we play our role as architect, teacher, and fellow explorer to the generations to come. The present generation has the gift of being able to inspire the joy of wondering and questioning from schoolchild to schoolchild. Mathew's book is just one tool in that arsenal, and his readers are encouraged to use this book to its fullest potential as one way to encourage future generations to their fullest potential.

Boston, MA, USA   *Debra Leahy*
July 2, 2012

Foreword 2
By Ishwar K. Puri

Chinese mythology describes a prehistory during which the Heavens and Earth were intermingled in an egg-shaped cloud. The Mossi of Africa believed in an initial phase devoid of matter and time, thus circadian rhythms. The Hindu Upanishads refer to a beginning when there was nothing. Such philosophical ruminations about our inception and origin have consumed us for millennia.

The American Hopis speak of four original couples. Each of these couples spoke a different language, but all four understood the others. Their progeny spread over the world and lived harmoniously. Over time, though, the groups separated, as they dwelt more on their differences than their similarities.

Such separations lead to dissolution. In 1854, President Franklin Pierce demanded of the Northwest Duwamish tribe that the U.S. government receive two million acres of their land. A resigned Chief Seattle responded, "No, we are two distinct races and must ever remain so. There is little in common between us." But he also said, "We *may* be brothers, after all. We shall see."

At an atomic and cosmic level, we are all one with our universe, which, in turn, is also part of us. Then, while yearning for an understanding of our existence, why do we embrace different answers, often stridently? Why do we separate from each other when we are able to reassemble and prosper?

Clearly, many forces can separate us from one another. However, since there are many more that bind us together, one can only hope that a common understanding of our inception will strengthen our bonds and fray our disagreements.

Modern science has led to a more thorough explanation about our origins. However, its pursuit has also led to many new questions. These delightful essays that describe our current understanding of the universe are therefore timely.

Intertwining cosmology and astronomy with philosophy, the themes are eloquently covered and provide a common understanding in an accessible manner for most readers. As a result, this book makes for a wonderful read.

As our scientific understanding advances, so does our society. What had previously seemed impossible becomes likely. It took about 1,400 years from the time of Ptolemy in AD 150 to Copernicus in the year 1543 to advance a geocentric view of the world to one that was heliocentric.

In the book, our author quotes Vera Rubin, whose work established that galaxies contain dark matter. "In a spiral galaxy, the ratio of dark-to-light matter is about a factor of ten. That's probably a good number for the ratio of our ignorance-to-knowledge. We're out of kindergarten, but only in about third grade."

Look what we've been able to do as a result of realizing that our Earth revolves around the Sun. Imagine then what we'll be capable of when we are able to more clearly understand the workings of our universe.

Blacksburg, VA, USA   *Ishwar K. Puri*
July 5, 2012

Preface

"I do not know what I may appear to the world, but to myself I seem to have been only like a boy playing on the seashore and diverting myself in now and then finding a smoother pebble or a prettier shell than ordinary, whilst the great ocean of truth lay all undiscovered before me."

– Isaac Newton

As a child, I grew up in the southern Indian state of Kerala, where the only thing that can come from the sky is the raindrops. However, the year 1979 changed that notion. One of the events I still remember vividly is the falling of Skylab scattering debris across the southern Indian Ocean and sparsely populated western Australia. Weeks before its descent from the orbit, where it had been in action for 6 years, there was mounting speculation over where the spacecraft would come down. My hometown newspaper and the radio had been constantly providing information about Skylab and the consequence of its imminent plunge. Even when people were discussing their gruesome fate in the event of it falling over our hometown, I had only one question in my mind: Why is it coming down?

The local priest in my hometown, as usual, offered a divine explanation linking the plunge of Skylab to the fulfillment of the scriptures and God's retribution for humankind's efforts to voyage into space or reach for the stars. He used this opportunity to exhort people to be more faithful, and embrace the holy wisdom that has been on a decline, a sign of apocalypse in his view. Although the neighboring Hindu temple had a special Puja[1] to divert the path of Skylab, not surprisingly the leftists in the state were enthused by the apparent fall of an American spacelab.

The only logical answer, which I partially remember now, came from my brother Abraham Mathew, who was an undergraduate student those days. He tried to explain the fierce competitive space programs Americans and Soviets were engaged in, and this was the result – Americans losing control of their spacelab. Truly, this was a convincing example of how the laws of nature can be characterized by the customs and beliefs of people,

---

[1] Puja is the act of showing reverence to a God, a spirit, or another aspect of the divine through invocations, prayers, songs, and rituals.

and how it can have a profound impact on the way science is perceived among the different segments of a society.

Nevertheless, it took me many years to fully recognize the fact that the force that pulled down Skylab is the same force that brought down a coconut tree nearby my home a couple of days before this event or, later on, to appreciate the omnipresent force of gravity that creates all the structures of the universe like a cosmic sculpture. Yet, the same force of gravity destroys the very same structures it once made in a never-ending process that has been going on since the beginning. But, this event was an enormous learning experience for me.

As I quoted Newton above, I am still a boy looking at the great ocean of the universe to which we all are linked in many ways. "We are made of star stuff," said Carl Sagan, and his influential words still resonate in my heart. When my thoughts ponder over the universe, it's a perfect moment of joy and contentment. However, deeper efforts are needed to know the universe, and that process helps us to know ourselves better.

This book is a collection of 12 essays on different topics that I have written in the past 2 years. This is my earnest effort to share with you the underpinnings of the magnificent cosmos where you and I are given a chance to exist briefly. I have avoided technical jargon and mathematical equations, except where absolutely necessary, to reach readers across the spectrum.

One of the fundamental questions I like to ask myself and in turn share with my readers throughout the book is the meaning of the laws of nature and our ability to comprehend them. To begin to understand, for example, that the force pulling on terrestrial objects is the same as the force that keeps the celestial structures is to understand an unimaginably powerful tool in science.

While modern science is able to answer some of our queries, it also poses many new questions. My approach was to intertwine cosmology and astronomy with philosophy and mythology, as these seemingly different schemes of thoughts provide a stimulating intellectual exercise.

I have to confess that the tenets of religion and mythology cannot fit into the rational and coherent framework of science; yet, the fundamental questions posed in any stage of human evolution are reflective of our curiosity at large.

Why do we seek answers or ask questions?

Some would argue that many cosmological studies transcend any practical purposes. True, but our species is distinct from any other because of our ability to ask questions and think beyond mere survival and procreation. Practical purposes change from time to time, not the fundamental science. We are here, it seems, to know our universe and thus ourselves better.

We live in an exciting time where our world is changing rapidly. Our scientific voyage is becoming more and more intriguing, where the minute particles such as the Higgs boson and the planets beyond our Solar System are no longer just pure dreams. Armed with imagination and curiosity, our species is on a journey that, I consider, is more enjoyable than the destination. This book is an attempt to join that great scientific expedition, and I hope you will be part of that journey along with me.

I believe that this book will mark a tiny footprint on the vast sand of space and time even as our sublime lives depart to become part of the grand cosmos.

I hope that you enjoy reading it.

Santhosh Mathew
July 2013

Acknowledgments

This book would never have been a reality without the help and support of many people. Apart from my own effort, the success of this book depends largely on the encouragement and guidance of many others. In that respect, a great debt of gratitude is due to many people who supported me throughout the process of writing this book.

I am indebted to my friends and colleagues for their valuable time and suggestions in preparing many topics. Many individuals from the community have inspired and encouraged me since I began writing, and I express my gratitude to all of them.

First and foremost, I would like to thank Maury Solomon, Editor, Springer, who relentlessly supported and helped me to complete this book. She oversaw every aspect of the publication of this work, which began the day I sent in the book proposal. The prompt responses to all my queries and the invaluable advice I received from her enabled me to keep the schedule, and I acknowledge her role as the most important force in shaping my words into a complete work.

My sincere appreciation goes to Prof. Abraham Loeb, chair of the astronomy department at Harvard University, who has graciously agreed to comment on my book. Professor Loeb has always been a true source of inspiration and support for me and devoted his time and expertise to meet all my requests on several occasions.

I take immense pleasure in thanking Ishwar K. Puri, professor of engineering science and mechanics and department head at Virginia Tech, for the brilliant suggestions and the foreword. Needless to say, Prof. Puri has always offered tremendous support and expressed interest in my research since I met him about 10 years ago.

I wish to express my deep sense of gratitude to Prof. David Morimoto, director of natural sciences division at Lesley University, Cambridge, for commenting on my writings. Professor Morimoto played the role of a mentor and academic guide from the beginning of my academic life in Massachusetts.

I would like to thank C.E. Larence (Carole Bugge), author and teacher, for offering comments on my essays. I must mention that the writing classes I have taken with her through the Gotham Writers' Workshop have truly improved my writing abilities.

Debra Leahy enthusiastically read the entire manuscript and wrote a detailed foreword to accompany this book. I would like to express my sincere thanks and gratitude to her for giving me such attention and time.

Special thanks to Achal Mehra, editor of *Little India* magazine, who was instrumental in publishing many of these essays. Achal offered meaningful suggestions and editorial helps throughout my writing career.

My sincere thanks to Malcolm Asadoorian, Dean of the School of Liberal Arts and Social Sciences, Regis College, for his support and encouragement during the process of completing this work.

Many thanks also go to other friends and colleagues for all their motivation and support to complete this book, including Sr. Barbara Loud, Cristina Squeff, Julia Benson, Upasana Kashyap, Laurie White, John Gostan, Barry Zaltman, Elizabeth McConnell, and Daniel Bielenin.

Thanks to my wife, Sumy, and son, Nathan, for having the patience with me for having taken on yet another challenge. Also, I would like to thank my parents and siblings who always supported and encouraged me in this endeavor.

Thank you to all who are drawn to read this book. Finally, I want to thank all those who provided support, talked things over, read, wrote, offered comments, and assisted in the editing, proofreading, and design.

I hope these essays will inspire you to learn about the unknown facets of the known universe and empower you to explore the meaning of existence.

Contents

About the Author

Santhosh Mathew, Ph.D., is a professor, researcher, and science writer. He has written extensively on scientific topics of popular interest. His regular science blog can be found at http://www.huffingtonpost.com/santhosh-mathew-phd. His writings on science and Eastern philosophy have been widely read. His research interests include developments in modern physics, mathematics, and cosmology.

Figures

1

Genesis of Genesis

The most beautiful and deepest experience a man can have is the sense of the mysterious. It is the underlying principle of religion as well as all serious endeavors in art and science. He who never had this experience seems to me, if not dead, then at least blind.

— Albert Einstein (1931)

THE BEGINNING OF THE BEGINNING

The mystery of the beginning is so profound that it occupies one of the primary places in science and religion. Thinking about the beginning of the "beginning" has always been appealing to human minds regardless of our inability to come up with definite and convincing answers for many facets of the oldest and most complex story – the story that is being written by the forces of nature on the fabric of the cosmos. We, the sentient beings, have just begun to comprehend the scripts of this marvelous story in a scientific fashion.

For cosmologists, the Big Bang is the story of the beginning. Surely it was big. The foundation of modern cosmology is based on the assumption that space and time originated from a unique event some 14 billion years ago, followed by the manifestation of everything else. Accordingly, the universe started with a bang – the Big Bang – from an unimaginably small but infinitely dense and hot state called singularity.

How do the cosmologists estimate the epoch of an event that is older than anything that we can imagine? The skeptics could argue that we don't even know about many events that happened much later.

S. Mathew, *Essays on the Frontiers of Modern Astrophysics and Cosmology*, Springer Praxis Books, DOI 10.1007/978-3-319-01887-4_1, © Springer International Publishing Switzerland 2014

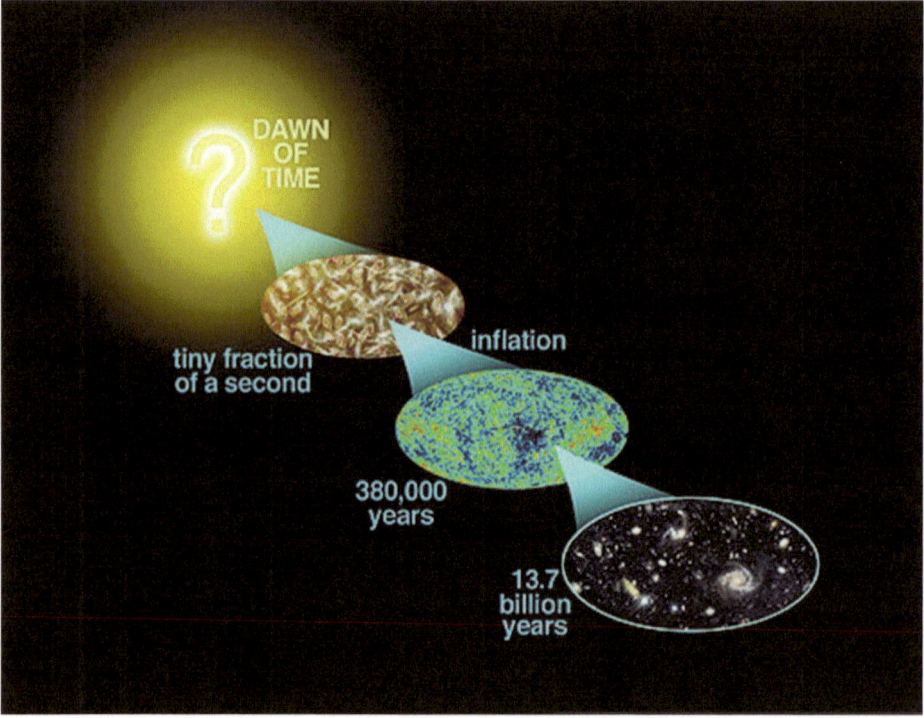

Figure 1.1. The beginning of the beginning. The materialization of space and time is scientifically considered as the beginning of our universe (Image credit: NASA).

Surprisingly, science has tools in its arsenal that can be applied to reach a conclusion. There are different sources from which the age of the universe can be estimated. These sources are:

- *The age of the chemical elements*
- *The age of the oldest star clusters*
- *The age of the oldest white dwarf stars (a white dwarf star is an object that is about as heavy as the Sun but only the radius of Earth)*

The publication of Edwin Hubble's article (1929), "A relation between distance and radial velocity among extra-galactic nebulae," was truly a turning point in understanding the universe. The Hubble Telescope is named after him.

In this paper, Hubble put forward the evidence for one of the greatest discoveries in twentieth-century science – the expanding universe. Hubble showed that galaxies recede from us in all directions, and more distant ones recede more rapidly in proportion to their distance.

This information also provides another way to estimate the age of the universe – from the cosmological model based on the Hubble constant (H_0) and the densities of matter and dark energy. This model-based age is currently $13.7 +/- 0.2$ Gyr (Gyr: gigayear, or 1 billion years).

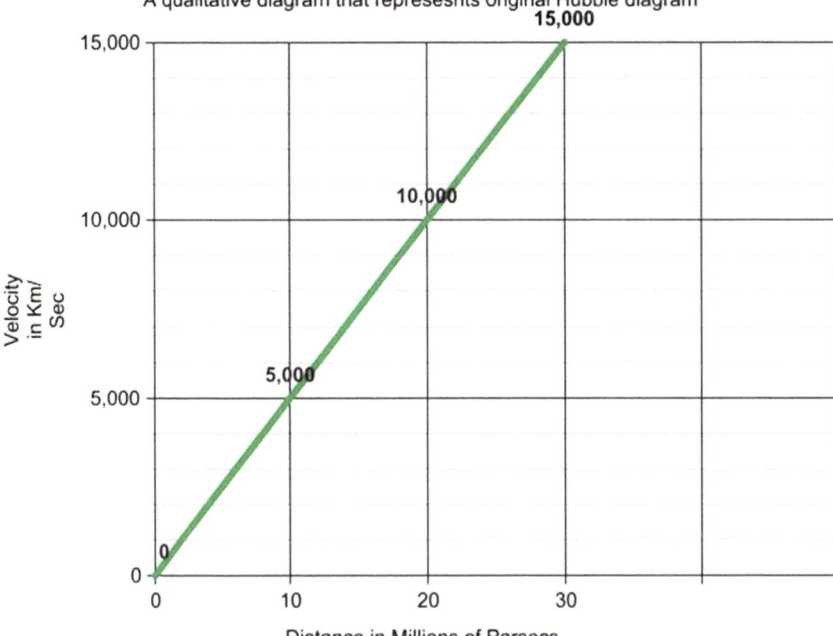

Figure 1.2. A graph that changed the universe. A replica of the original Hubble diagram from a 1929 paper, which is simply a plot of galaxy velocity versus galaxy distance (a megaparsec is about 3.09×10^{22} m). A revolutionary discovery coming from a seemingly simple graph that represents the observational data of Hubble.

It's hard to believe how such a simple mathematical relationship became a cornerstone of the study of the universe. However, the truths are simple and can be explained to anyone!

Mathematically, the slope of the line (Fig. 1.2) can be estimated by using the formula $H_0 = \text{Velocity/Distance} = 1/\text{Time}$. The time value that H_0 represents is the time it took for the universe to expand to its current size. Another way of saying this is that it is a rough estimate for the age of the universe. As you can see, this is a pretty simple algebraic equation but a hugely significant one!

We can deduce that the slope of the line in the above diagram is same as the Hubble constant (H_0). Hubble's Law says that velocity $= H_0 \times$ distance. As we know, time = distance/velocity, there is a natural time associated with the Hubble expansion. We can calculate, Time $=$ distance/velocity $=$ distance/($H_0 \times$ distance) $= 1/H_0$.

The time value that H_0 represents is the time it took for the universe to expand to its current size from its origin. Another way of saying is that it is a rough estimate for the age of the universe. As you can see, this is a pretty simple algebraic equation but a hugely significant one!

Latest work (Garnavich, P. M. et al.) and observations with the Hubble Space Telescope gives $H_0 = 72$ km/s/megaparsec. The calculation with Hubble constant of 72 km/s/Mpc is shown below.

This exercise is intended to provide an idea of converting Hubble's constant to an age of the universe. It is important to understand the connection between this constant and the age of the universe before we inquire about the social context of such a finding.

Though different estimates show that the Hubble constant H_0 varies from 50 to 100 km s^{-1} Mpc^{-1}, we would consider a value of 72 km s^{-1} Mpc^{-1}

$1\ pc = 3.09 \times 10^{16}$ m

$1\ Mpc = 10^6$ parsec $= 3.09 \times 10^{22}$ m

The age of the universe is:

$$t_{0=} \cfrac{1}{\left[\cfrac{km\ /\ sec}{Mpc} \right]} = \cfrac{1}{72 \left[\cfrac{km\ /\ sec}{3.09 \times 10^{19}\ km} \right]}$$

$$= \cfrac{1}{2.33 \times 10^{-18}\ \cfrac{1}{sec}}$$

$$= 4.29 \times 10^{17}\ s$$

1 year $= 365$ days/year $\times 24$ h/day $\times 60$ min/h $\times 60$ s/min $= 3.16 \times 107$ s

Converting seconds to year $4.29 \times 10^{17} \times \cfrac{1}{3.16 \times 10^7\ sec}$

$$= 13.5 \times 10^{12}\ \text{years}$$
$$= 13.5\ \text{billion years}$$

One of the most precise measurements of the age of the universe, fixed at around 13.7 billion years, was made by NASA's Wilkinson Microwave Anisotropy Probe (WMAP). More details about this will be discussed later.

MORE ON EDWIN HUBBLE

Edwin Powell Hubble was born in the small town of Marshfield, Missouri, on November 29, 1889. He studied law at the University of Chicago and later at Oxford, but he managed to keep his interest in science and mathematics. Later on, he entered the graduate program at the University of Chicago with the intention of becoming a professional astronomer and completed his doctoral thesis entitled "Photographic Investigations of Faint Nebulae."

During World War I, Hubble enlisted in the U. S. army, and after the end of the war he was able to renew studies in astronomy. Later on, he was offered a position at the Mount Wilson Observatory in Pasadena, California. This was where he performed his ground-breaking observations in modern astronomy. Mount Wilson was the center of observational work in those days and housed the 100-in. Hooker Telescope, then the most powerful on Earth.

Hubble's work proved the existence of galaxies outside of our own Milky Way Galaxy, which was thought to be the entire universe until then. He also classified the various galaxies he observed based on their content, distance and brightness. The famous Hubble's law, discussed earlier, was formulated based on these observations.

Astronomers even today use the system developed by Hubble known as "Hubble tuning fork." Hubble separated the galaxies into two general categories: elliptical galaxies and spiral galaxies. Elliptical galaxies are shaped like ellipses, and spiral galaxies, as the name suggests, resembles spirals. Although the theory of galaxy evolution proposed by Hubble is wrong, the diagram was the first attempt at such a classification.

Hubble could easily be awarded the title of the greatest observational astronomer of modern times. Hubble died in 1953, leaving "one of the great intellectual revolutions in the twentieth century" in the words of Stephen Hawking.

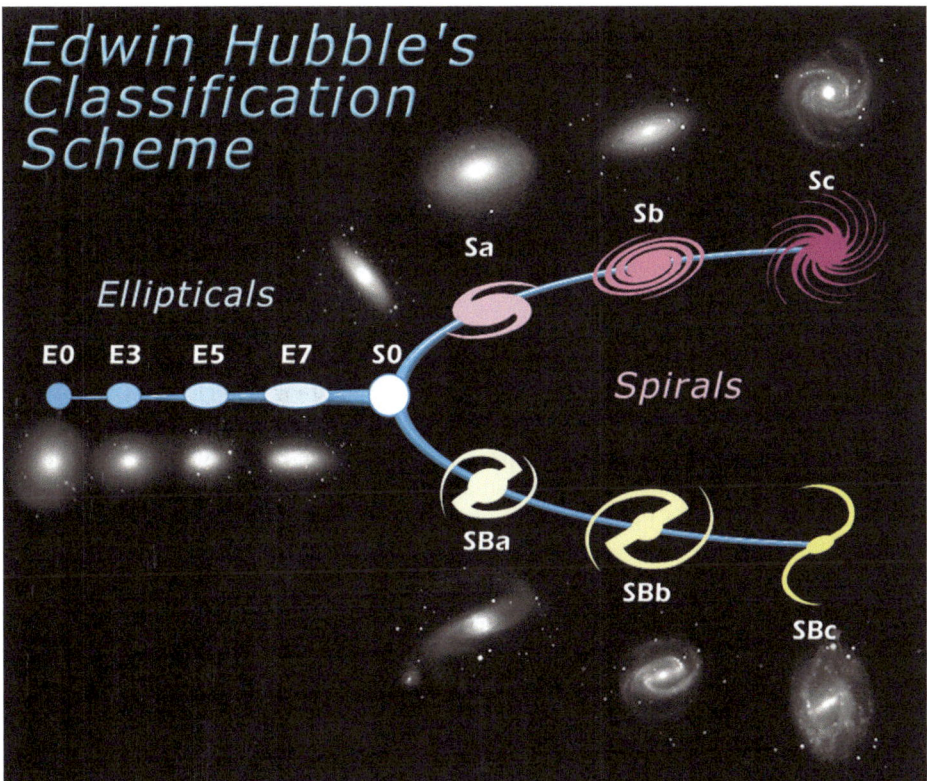

Figure 1.3. Hubble tuning fork (Image credit: NASA/The Space Telescope Science Institute).

COSMIC HISTORY OR ILLUSION?

Why do we care about the age of the universe? Should we be spending money to measure the age of the universe? Why are such questions important? Why has the origin of the universe fascinated humans for so long?

The simple answer is that we are inquisitive by nature. In the early seventeenth century, when the idea of telescopes was introduced for the first time, they were meant to be spyglasses – a tool that would refract or reflect light to take us beyond where our own eyes can. However, they do truly reflect the intrusive nature of the humans who use them. These awesome instruments shaped our picture of the universe by stretching our sense of sight far beyond the realm of our forebears' imaginations.

Imagine that anyone seeing Galileo looking through this spy glasses could easily have wondered or even been perturbed about the futility of such an act! What benefits might it have brought in the seventeenth century for someone looking to the sky rather than looking for approaching war ships or spying the enemy posts? Nevertheless, such random acts of audacity shaped our civilization in numerous ways.

The previously discussed method of calculating the age of the universe many not be the perfect or final answer in the quest to know the history. The Hubble constant is a measure of the current expansion rate of the universe. It is possible that the universe had a different rate of expansion in the past. This implies that Hubble's constant was different in the past and it could vary in the future. In other words, it's not a perfect constant. This in turn would tell us that the age of the universe we accept may not be the true age of the universe. There are different scenarios one has to consider, such as flat, open or closed geometrical shape of the universe, when using the currently known value of Hubble's constant.

Yes, the universe confronts us with an age issue. This is exactly the reason science exists and scientific history teaches us that the scientific path is always not just an easy, smooth road. And no doubt, even more questions will be raised that we have not yet thought to ask.

As of March 2013, the European Space Agency's Planck mission came up with the most detailed ever map of the oldest light shining through the universe. Based on the initial 15.5 months of data from Planck, we could see the perfect picture of a universe when it was just 380,000 years old. This not only attests to the accuracy of current cosmological models, but we were also able to place the age of the universe at what is now believed to be an even better estimate: 13.82 billion years.

The estimation of the age of the universe serves as an excellent example of how science evolves through a combination of independent research, development of theory, discussion, and correction of error. The greatest friend of science is truth, and it must be so.

Let's go back to the concept of singularity, the state from which the baby universe sprung into action. Even though we are able to estimate the age of the universe, it is well known that the accepted laws of physics, such as gravity and relativity, break down in the state of singularity.

We don't have many details about this primordial age of the universe. If we depend purely on scientific terminology, we could say that quantum fluctuations caused the Big Bang and created the universe. Quantum fluctuation is the temporary appearance of

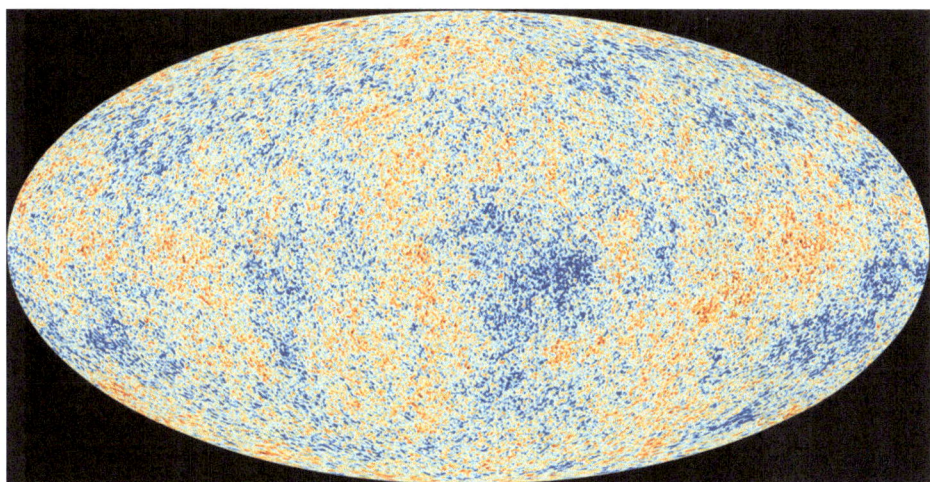

Figure 1.4. The cosmic microwave background (CMB) as observed by Planck. The CMB is a snapshot of the oldest light in our universe, imprinted on the sky when the universe was just 380,000 years old. It shows tiny temperature fluctuations that correspond to regions of slightly different densities, representing the seeds of all future structure – the stars and galaxies of today (Image credit: ESA and the Planck Collaboration).

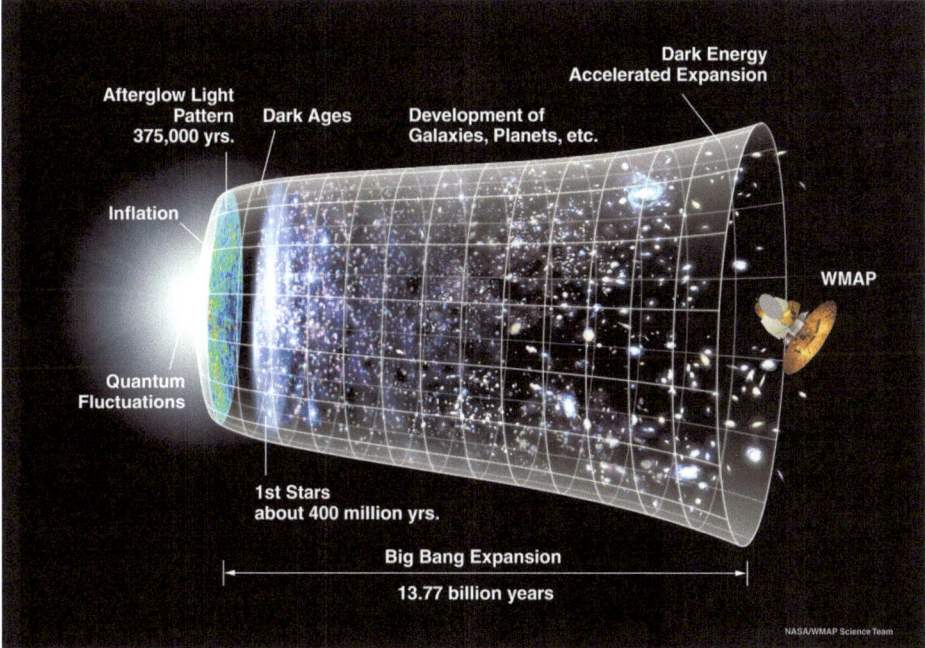

Figure 1.5. This diagram traces the 13.7-billion-year history of the universe from the quantum scale to the formation of stars, galaxies, planets, and WMAP (Image credit: NASA/WMAP science team).

energetic particles out of nothing, as allowed by the uncertainty principle. It is synonymous with vacuum fluctuation. One could simply ask what caused the quantum fluctuations. And this can go on and on…

THE BEAUTY OF VASTNESS

To get a feel of the vastness of the universe, think of our cosmic neighborhood. The Sun, the star at the center of our Solar System, is almost 93 million miles away. It takes light from the Sun 8 minutes to reach us, so the Sun we see is actually the way it was 8 minutes earlier. The Sun dominates our Solar System, accounting for 99.86 % of its mass. Approximately 1.3 million Earths could fit inside the Sun, whose diameter of 870,000 miles is 110 times that of Earth.

However, even the Sun and our Solar System are a faint bit of dust in the vastness of the universe, which is infinite, so to say, in volume, populated with billions of galaxies, each with billions or more stars in the observable universe alone.

Distances in the universe are so vast that astronomers represent them in light-years, equivalent to 5.9 trillion miles, the distance covered by light, the fastest permissible speed, in 1 year. A typical galaxy is 30,000 light-years in diameter and is separated from its nearest neighbor by 3 million light-years. Our galaxy, the Milky Way, is 100,000 light-years across, and the closest galaxy to us is 2.5 million light-years away. It would take us 100,000 years traveling at the speed of light, 186,000 miles per second, to traverse our galaxy edge to edge.

These enormous distances are not the only mind-bending thoughts in cosmology. Equally important are the vast number of galaxies that inhabit our known universe – aptly called the galaxy zoo. How many galaxies are in the universe? The most current estimate put the number somewhere between 100 and 200 billion. As we know, each of them contains billions of stars and the planets they carry with them. They come in all different configurations. There are large, small, old, young, red, blue, regular, irregular, luminous, faint – whatever size or shape you could imagine.

These gigantic beasts fill the universe with matter and light and yet they harbor the secrets of the cosmos in their womb in the form of black holes. Most galaxies have been known to be hiding supermassive black holes at their center, in addition to numerous black holes that reside throughout the galaxies.

Super massive black holes (SMBHs) may be the engines that drive the galaxies across cosmic time. Thus it is imperative that for having a complete picture of galaxy formation, these processes of galaxy and black hole evolution can no longer be regarded as separate, as was the case until about 10 years ago, but need to be studied in conjunction. Moreover, there is now an emerging consensus from dynamical observations that SMBHs at the center of the local massive and bulge-dominated galaxies are tightly correlated with the velocity dispersion and masses of their stellar hosts (Bongiorno et al. 2012).

Galaxies collide and merge as they continue their cosmic ballet. In essence, they evolve as any other entities in the universe. However, these gravitationally bound systems do move away from each other as observed and recorded for the first time by Edwin Hubble (1929). Remember the Hubble constant (H_0) and the age of the universe we deduced in previous pages. The basis of these calculations are the motions of the galaxies, now known as the accelerating universe.

THE DOPPLER EFFECT AND REDSHIFT

The apparent change in the pitch (or frequency) of sound is called Doppler shift, named after the Austrian physicist Christian Doppler, who proposed it in 1842. There are many everyday examples of the Doppler effect that we experience – for example, the changing pitch of police and ambulance sirens, or train whistles and racing car engines as they pass by. This very basic principle is used in a variety of tools such as radar guns used by police to check for speeding vehicles and meteorologists use it to record the weather events. By measuring the change in frequency of the waves that reflect from an object, one can deduce the velocity of the object.

Analogous to sound, the frequency of light from distant stars and galaxies can be shifted in very similar fashion. The redshift of a distant galaxy or quasar is easily measured by comparing its spectrum with a reference laboratory spectrum. Atomic emission and absorption lines occur at well-known wavelengths. By measuring the location of these lines in astronomical spectra, astronomers can determine the redshift of the receding sources.

As an object moving toward an observer, its frequency appears to be increased, and the frequency from the object moving away from an observer appears to be decreased. These are known as blueshift and redshift, respectively. As in the case of sound waves these shifts exhibited by astronomical objects indicate their motions with respect to the observer.

Measuring the redshifts of the galaxies and stars has become an important tool in astronomy. It is to be remembered that these shifts observed in distant objects are the result of the expanding universe. In other words, objects that are stationary in space could exhibit redshift if the intervening space itself is expanding.

As we have seen, the basic premise of the Big Bang theory is that the universe is expanding due to the fact that galaxies are receding from each other. This velocity can be expressed as a function of its distance by the linear relationship $v = H_0 d$, where H_0 is the Hubble constant and d is the distance of the object from Earth.

So to determine an object's distance, we only need to know its velocity. Velocity is measurable by utilizing the Doppler shift. By taking the spectrum of a distant object, such as a galaxy, astronomers can see a shift in the lines of its spectrum and from this shift determine its velocity (Fig. 1.6). Putting this velocity into the Hubble equation, they determine the distance.

Let's look at an example here:

Suppose the redshift (z) of an object measured from the wavelength shift of its light is 0.15. At low velocity (less than about 15 % the speed of light) the z-value of a redshifted object is almost equal to the recessional velocity of the object expressed as a fraction of the speed of light. This means that if $z = 0.15$ then the object producing this light has a recessional velocity of $v/c = 0.15$ or $v = 0.15c$, where c is the speed of light (300,000 km/s).

So in this case the recessional velocity of the galaxy $= 45,000$ km/s.

We need to emphasize that the formula $v = c z$ is accurate only when z is small compared to $z = 1.0$. For larger redshifts, a more sophisticated equation that accounts for effects predicted by Einstein's Special Theory of Relativity must be used. Also remember the galaxies are not just moving through space, but they are being carried along by space as it expands.

Now, if we want to calculate the distance of this object, we could go back to the equation $v = H_0 d$. Assuming the Hubble constant 72, D $= 625$ Mpc, which is over 2 billion light years.

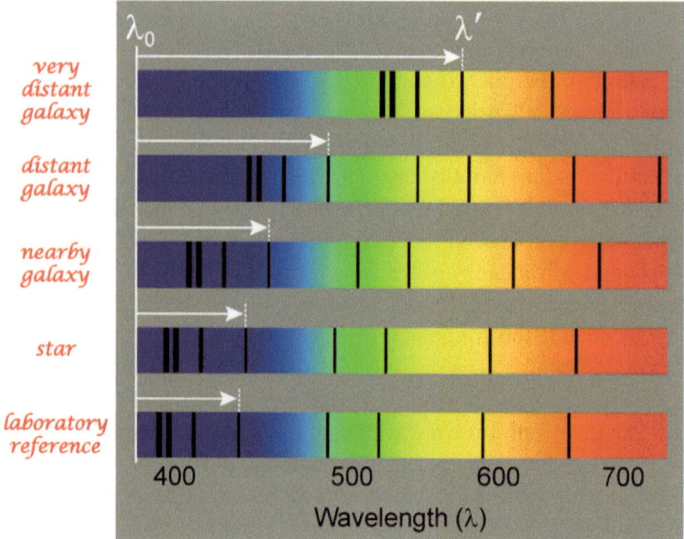

Figure 1.6. Redshift and blueshift of light in a schematic diagram. The radiation from most galaxies shows that the light is redshifted, indicating they are moving away from the observer. Hubble found that spectral lines in the light from distant galaxies are shifted toward the *red* (longer wavelength) end of the electromagnetic spectrum (Image courtesy of NASA/JPL-Caltech).

The astronomical distances and time scales are incomprehensible in terms of their magnitude; however, the human capacity to comprehend the incomprehensible begins with the baby steps of scientific inquiry. There was a time, not so long ago, we thought that the universe was a single galaxy and planets are confined to our own Solar System.

"The history of astronomy is a history of receding horizons," said Edwin Hubble (1982), who not only excelled as an intellectual heavyweight but a heavyweight boxer as well. The history of scientific discovery is a history of the unexpected, and many such unexpected are yet to come, and one of those unexpected was quasars.

QUASARS, THE BEACONS OF THE UNIVERSE

When we talk about the unexpected in astronomy, we cannot fail to talk about quasars, which stand for quasi-stellar radio sources. Quasars are described as the accreting discs around massive black holes at the center of distant and young galaxies. In fact when they were discovered, some astronomers initially thought of them as the signs of intelligent life in the universe, because they beam their light across the universe with a definite periodic precision.

Many astronomers believe that quasars are the most distant and brightest objects yet detected in the universe. Astronomers were stunned by the enormous amounts of energy the quasars give off – trillion times more than that of the Sun! Quasars are believed to produce their energy from massive black holes in the center of the galaxies in which the quasars are located. They outshine all other stars in the host galaxy. The early speculation about them being stars was ruled out on the basis of the energy they emit. In fact, they have nothing in common with stars.

In previous sections we have seen how redshift is associated with distances. The quasars have very large redshifts, and that indicates they are at a greater distance. Looking at such large distances in the universe is equivalent to looking back in time because of the finite speed of light. Thus, the observation of quasars at large distances and their scarcity nearby implies that they were much more common in the early universe than they are now.

Maarten Schmidt, a Caltech astronomer working at Mt. Palomar Observatory, was credited with discovering the first quasar in 1963. Since then hundreds of quasars have been identified. However, astronomers are still struggling to understand how quasars are formed.

The study of quasars point to some astonishing and interesting facts. The realization that the early universe had many more quasars than today has deeper implications for astronomy. It might help astronomers to decode the early history of galactic formation. It also shows the violent history of the universe as these objects feed black holes. The quasars are not only the beacons of the universe but also serve as a guide that leads to the extreme past of our cosmos.

Figure 1.7. This artist's concept illustrates a quasar, feeding a black hole. This quasar contains a black hole with a mass several times that of the Sun, and is so far from Earth that its light takes a billion years to reach us (Image credit NASA/JPL Caltech).

Although quasars teach us the history of the universe we also realize that they themselves are history. As mentioned earlier, none of them exists in our cosmic neighborhood. In other words, quasars were not created during our current era of the universe, but during an ancient era. This also implies that the universe was quite a different place in the past (billions of years ago). In addition, it can be concluded that the galaxies we see around us now may have been quasars in the distant past. This suggests that even our own Milky Way Galaxy may have been a quasar-like galaxy long ago.

The warm heart of the Milky Way that hides a super massive black hole now might have once been a boiling hotspot in the past. Despite the fact that it has been 50 years since the discovery of the first quasar, quasars still continue to feed us with new pieces of information that will help us to solve some of the great cosmological puzzles we are dealing with.

PRE BIG BANG

Although we do not completely understand the pre-Big Bang era, the events that followed are well comprehended. After the Big Bang, the newly born universe inflated and continued to expand. Basically, from the sheer emptiness emerged our magnificent cosmos! Scientific observations support such a scenario, but no one knows what happened at the time of the birth, let alone before the event.

As a result, usually cosmologists shy away from the question, "What happened before the Big Bang?" dismissing it as a purely philosophical question, as the Big Bang itself represents the origin of everything we know, even the absolute emptiness that we associate with nothing. Hence, it doesn't make sense to most of them to talk about something before that.

One could argue, of course, that this theory of scientific creation is not much different from the mythological or religious stories of creation except that it excludes a creator by transferring that role to the laws or forces of nature. In fact, some experts question the uniqueness of this event and claim they have found evidence that the Big Bang occurred several times and that the symphony of the never-ending cosmic origin is resonating everywhere.

Oxford physicist Roger Penrose is one of the few physicists to confront the pre-Big Bang-era issues. In a recent preprint paper, Penrose and another physicist, Gurzadyan (2010), claim that they now have observational evidence of previous Big Bangs. Their conformal cyclic cosmology (CCC) posits the existence of an aeon[1] preceding our own Big Bang. The cyclic nature of the universe was hypothesized much earlier by different researchers, but this is the first time that empirical evidence has been put forward. Penrose and Gurzadyan argue that the laws of nature may evolve with time, but their theory precludes the need to introduce a theoretical beginning to the universe, as we currently presume. They use the term "aeon" to describe the period from our Big Bang until the remote future.

Penrose explained to the BBC, "I claim that this aeon is one of a succession of such things, where the remote future of the previous aeons somehow becomes the Big Bang of our aeon."

---

[1] An immeasurably or indefinitely long period of time.

In essence, the Big Bang is bound to happen again and again, creating new universes and recycling old ones. Simply put, anything that can happen is bound to happen infinitely.

The Big Bang, including its name, has always been a subject of debate. The name itself is a misnomer, as there wasn't any bang in the first place. A bang (sound) needs to have a vibrating particle to generate it, and it requires a medium for the vibrations to propagate. Since the Big Bang is the beginning of anything we can imagine, it is logical to assume that this moment of creation was in perfect silence.

This theory was one of several theories, trying to account for the origin of the universe, floating around in scientific circles in the 1930s. Georges Lemaître, a Belgian Catholic priest, astronomer, and physics professor, applying Einstein's theory of relativity, proposed the Big Bang as the possible origin of the universe. It essentially suggested that the presently expanding universe must have been small in the past, smaller than an atom, where all the known forces coexisted. Thereafter came the "explosion of this cosmic egg" that created the universe, which has been expanding since its birth, about 13.7 billion years ago.

Fred Hoyle, Hermann Bondi, and Thomas Gold developed an alternative mathematical model of the universe that did not start in a colossal expansion. In their version of the theory, the universe had no beginning or end, and it must have looked the same always. According to this theory, there never had been a "Big Bang" – a phrase that Hoyle invented in 1950 in a series of radio talks on astronomy for the BBC. He intended the nickname as a pejorative.

Although Hoyle coined the term Big Bang to ridicule the theory, the name stuck, and Lemaître's theory has come to be known as the Big Bang theory. American astronomer Edwin Hubble's discovery that the galaxies are flying apart in an apparently expanding universe provided major support for the Big Bang theory. Eventually Hoyle's Steady State theory faded from the scientific community like the receding galaxies. The death knell for the theory sounded when radio astronomers Arno Penzias and Robert Wilson (1965) discovered the cosmic microwave background, the leftover radiation from the Big Bang.

However, as one might expect, the Big Bang theory doesn't address the question of what created the Big Bang. As mentioned earlier, cosmologists believe that complete quantum chaos must have existed at an early moment of the Big Bang, and a quantum fluctuation might have caused the bang. Such explanations lack details and are no better than mythological stories of origin, some say.

Nonetheless, the theory is the standard framework within which most cosmologists now operate. Its position is akin to that of the theory of evolution in biology. But some critics say this theory is propagating the Judeo-Christian concept of creation, and science has sacrificed its soul to theology. The criticism stems from the fact that the theory confirms the theological notion of creation – that the universe has a definite beginning and is created with a fixed design.

CREATION MYTHS

At this junction, we will revisit the topic of eastern mysticism and its alleged and well-discussed connections to modern physics. It's a topic that has been brought into popular consciousness in a big way through bestsellers such as *The Tao of Physics* by Fritjof Capra (1979).

The intention of introducing that discussion is to show how creation myths are an integral part of all ancient cultures and their philosophies.

Although the monotheistic religions depend on a universe with linear progression since origin, ancient eastern philosophies such as Hinduism propose a recurring nature of the universe. Accordingly, the universe has no beginning or ending, but is comprised of never-ending cycles. Even the belief of birth and rebirth reflects that philosophy.

In his popular television series *Cosmos,* renowned astronomer Carl Sagan reflected on the parallels between modern cosmology and Hindu philosophy, noting:

> Hindu religion is the only one of the world's great faiths dedicated to the idea that the cosmos itself undergoes an immense, indeed an infinite, number of deaths and rebirths. It is the only religion in which the time scales correspond, no doubt by accident, to those of modern scientific cosmology. There is the deep and the appealing notion that the universe is but the dream of the God who after a hundred Brahma years dissolves himself into a dreamless sleep and the universe dissolves with him until after another Brahma century, when he starts recompos[ing] himself and begins again the dream, the great cosmic lotus dream. Meanwhile, elsewhere there are an infinite number of other universes, each with its own God dreaming the cosmic dream.

As Sagan pointed out, such similarities could be accidental in nature, as some like to think, or ancient wisdom as expressed by the sages, as some others like to believe.

Figure 1.8. Every religion has story for the creation. But scientific reasoning is by all means the most acceptable one (Image credit: CC by the York Project on Wikimedia commons).

The moral fiber of science that we inherit, however, does not tempt one to propose a definite conclusion on such thoughts. Yet, it is interesting to explore such ideas.

For instance, as cosmologists struggle to generate a consistent model about the pre-Big Bang era, the creation myths in Hindu cosmology, which describe Brahma,[2] the creator, bringing the universe into being through his thoughts, might offer some clues. The universe survives four different yugas within a maha yuga. At the end of this period, the universe undergoes a dark age, which in turn gives way to a golden age, and this cycle repeats. After the lifetime of Brahma, everything disappears into the Supreme Being, and after an unimaginable period of time, a new universe and new Brahma emerge. It seems that even the Gods follow the laws of nature.

The metaphorical similarities between the newly proposed cyclic universe theory and Hindu cosmology are astounding. Hindu cosmology speaks about the universe being created, destroyed, and recreated in an eternally repetitive series of cycles. Brahma creates a new universe that would eventually be destroyed to begin another universe. According to Hindu cosmology, Brahma himself undergoes destruction after a 100 Brahma years – 311 trillion years in the human scale – to pave the way for a new Brahma and thus a new universe. Gods are also subject to the laws of nature, and their lives occupy the cosmos in time scales that are comparable to the astronomical scales in such an oscillating universe. As some suggest, Kalpa, which is a day in the life of Brahma, is 4.32 billion years, the estimated age of Earth.

Although there are different accounts of creation in Vedic[3] collections, all of them clearly point to the self-repeating aspect of cosmic origins. It is comforting that, although these narratives lack scientific underpinnings and are thus not acceptable in the scientific realm, a cyclical model of the Big Bang will answer or at least negate many challenges the Big Bang model now faces.

However remarkable these similarities are, there are also significant differences between Vedic doctrine and modern cosmology. The Sun's diameter (32,500 miles, according to Surya Siddhanta, one of the oldest astronomical texts written in Sanskrit) is far from the modern-day value of 870,000 miles. Modern science postulates that organic molecules (matter) evolved to become life and consciousness, which is fundamentally different from Hindu philosophy's approach that they cannot evolve from matter unless matter has the inherent potential to provide life and consciousness. Furthermore, the Big Bang model proposes that the mass-energy before the universe came into being was concentrated at a single point. The Vedas, by contrast, assert that in the beginning there was no mass-energy, and it was a complete void. Even the cyclic model of the universe, while bearing some similarity to Hindu cosmology, is different. The ancient model says the cycles are independent, but modern cyclic theory assumes dependence between cycles.

The universe, it seems, has a special way of recording the birth and death of everything. The birth and death of stars are imprinted on the fabric of the cosmos. Fossils bear the indisputable evidence of the foregone era of plants and animals. How about the universe? In fact, the evidence of the "oldest birth" is everywhere.

---

[2] Brahma is the God of creation in Hindu mythology.

[3] Vedas are the religious texts considered to be the foundation of Hinduism.

EVIDENCE FOR THE BIG BANG

In the mid-1960s two researchers, Arno Penzias and Robert Wilson, working at Bell Labs in New Jersey, detected leftover, cooled down radiation from the early universe by carefully scanning the sky with a device called the Holmdel horn antenna. Their discovery is important evidence in support of the Big Bang theory and won them the Nobel Prize in 1978.

The discovery of the cosmic microwave background radiation (CMB) in 1964 becomes very significant in this regard. In fact, it secured the position of the Big Bang as the most acceptable cosmological theory to date and turned the subject from purely theoretical science to the most exciting scientific quest of our generation. The presence of this radiation can be detected everywhere, as Penziaz and Wilson showed, but it is not visible to the naked eye. If we could see microwaves with our eyes, the entire sky would glow with a brightness that is astonishingly uniform in every direction. This uniform radiation is the remnant heat from the Big Bang. It would be impossible for any other source to produce such uniform radiation in all directions.

NASA's Wilkinson Microwave Anisotropy Probe (WMAP), launched in June 2001, has mapped the CMB radiation. This is the oldest light or radiation in the universe, which emerged when the universe was about 380,000 years old. Currently, it remains in the microwave frequency range of the electromagnetic spectrum. Indeed, this radiation fills even our own living rooms. A small part of the noise one detects in the television or radio is from the CMB. Evidence of the Big Bang is always around us!

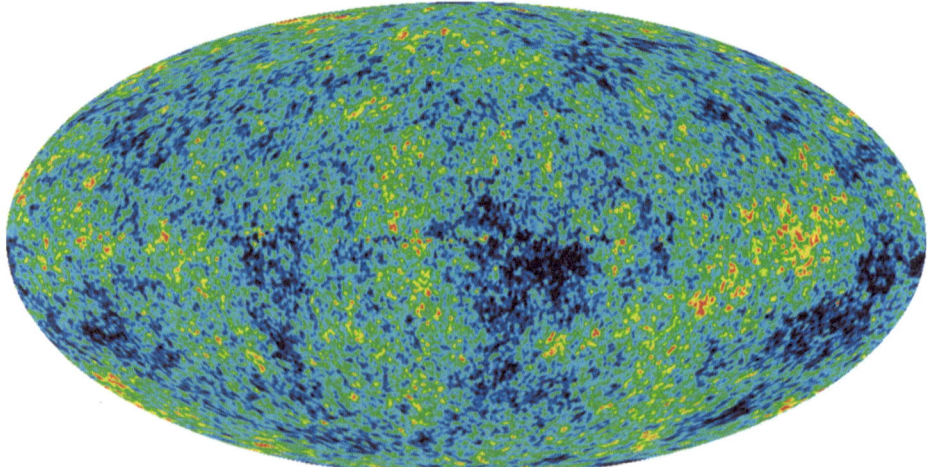

Figure 1.9. The detailed, all-sky picture of the infant universe created from 9 years of WMAP data. The image reveals 13.77-billion-year-old temperature fluctuations (shown as color differences) that correspond to the seeds that grew to become the galaxies (Image credit: NASA/WMAP Science Team-The Infant Universe).

Penrose suggests that analysis of this cosmic microwave background would show echoes of previous Big Bang-like events. The events appear as "rings" around galaxy clusters. Like the tree rings that represent the growth and age of a tree, these rings indicate the cycles of the grand cosmic tree. Accordingly, the concentric rings are caused by the clash of super massive black holes in earlier versions of our universe and imprinted ripples of smog of microwave radiation.

Many questions remain unanswered, and even researchers acknowledge it is not yet time to celebrate. Perhaps the Big Bang did not happen exactly as we envision it currently. The Planck satellite, currently in action, has just provided more sensitive data in the recent months. That would provide in-depth analysis of the cosmic birth and may force us to rewrite its history yet again.

The Planck space mission, an ESA mission with NASA contributions, has released the most accurate and detailed map ever made of the oldest light in the universe, revealing new information about its age, contents and origins. Scientists are currently analyzing the data for more details.

Overall, the findings of the Planck mission support the earlier revelations by WMAP with minor corrections. The latest results suggest that the universe is expanding more slowly than scientists thought, and is 13.8 billion years old, 100 million years older than previous estimates. The data also show there is less dark energy and more matter, both normal and dark matter, in the universe than previously known.

Why are such investigations relevant? Well, the fact is that we need a story and a foundation to link to our observations, because our innate nature tells us that cause precedes effect – though that could be wrong. As demonstrated by whatever awe one experiences at the vast expanse of the night sky, it is difficult for the human mind to comprehend the possibilities of the universe.

Cosmologists still believe our universe originated about 14 billion years ago with a Big Bang, although newer oscillating theories of the universe postulate that the Big Bang was only the most recent recreation of the universe, which has been through countless cycles of births and destruction over trillions of years.

The Big Bang theory explained the abundances of the most common elements in the universe in a groundwork laid out by Alpher et al. (1948). Although it still faces formidable challenges to meet the standards of science, it is the best prevailing model for the universe's origin. This is where our legacy began long before our Solar System or Earth was formed – let alone life.

Profound truths spring from imagination and observation, though they may appear nonsensical to begin with. Going back to a question we briefly introduced earlier, what is the value of cosmological research? Is it the responsibility of science that it needs to have a social impact on every venture it undertakes? Many would say no to this.

As physicist and Nobel laureate Richard Feynman (1970) said in his lecture, "The Value of Science," enjoyment of science is as important as anything else. He once quoted a Buddhist proverb: "To every man is given the key to the gates of heaven; the same key opens the gates of hell."

REFERENCES

Alpher, R. A., Bethe, H., & Gamow, G. (1948). The origin of chemical elements. *Physical Review, 73*(7), 803–804.

Bondi, H., & Gold, T. (1948). The steady-state theory of the expanding universe. *Monthly Notices of the Royal Astronomical Society, 108*, 252.

Bondi, H., Gold, T., & Hoyle, F. (1995). Origins of steady-state theory. *Nature, 373*, 10.

Bongiorno, A., et al. (2012). Seeking for the leading actor on the cosmic stage: Galaxies versus supermassive black holes. *Advances in Astronomy*, [Online]. 3 pages. http://www.hindawi.com/journals/aa/2012/625126/cta/. Accessed 11 Mar 2013.

Capra, F. (1979). *The Tao of physics. An exploration of the parallels between modern physics and eastern mysticism*. London: Fontana/Collins.

Einstein, A. (1931). The world as I see it. *Forum and Century, 84*, 193–194.

Feynman, R. P. (1970). The Feynman lectures on physics (3 Vols.) (Set v). Edition. Reading: Addison Wesley Longman.

Garnavich, P. M. (1998). Supernova limits on the cosmic equation of state. *The Astrophysical Journal, 509*, 74–79.

Gurzadyan, V. G., & Penrose, R. (2010). Concentric circles in WMAP data may provide evidence of violent pre-big-bang activity. arxiv.org 1. http://arxiv.org/abs/1011.3706. Accessed 9 Mar 2011.

Hubble, E. (1929). A relation between distance and radial velocity among extra-galactic nebulae. *Proceedings of the National Academy of Sciences of the United States of America, 15*(3), 168–173.

Hubble, E. (1982). *The realm of the nebulae* (The Silliman memorial lectures series). New Haven: Yale University Press. Edition.

Kirshner, R. P. (2013). Hubble's diagram and cosmic expansion. [ONLINE]. http://www.pnas.org/content/101/1/8.full. Accessed 14 May 2012.

Lemaître, G. (1931). Expansion of the universe, the expanding universe. *Monthly Notices of the Royal Astronomical Society, 91*, 490–501.

Loeb, A. (1998). Direct measurement of cosmological parameters from the cosmic deceleration of extragalactic objects. *The Astrophysical Journal Letters, 499*(2), L111–L114.

Maarten Schmidt Discovers Quasars | Everyday Cosmology (1963). [ONLINE]. http://cosmology.carnegiescience.edu/timeline/1963. Accessed 16 May 2013.

Manas: Religious texts of India, Puranas. (2013). [ONLINE]. http://www.sscnet.ucla.edu/southasia/Religions/texts/Puranas.html. Accessed 24 Mar 2013.

National Aeronautics and Space Administration. (2004). WMAP-age of the universe. Wilkinson Microwave Anisotropy Probe (WMAP). http://map.gsfc.nasa.gov/universe/uni_age.html. Accessed 8 July 2012.

Penrose, R. (2011). *Cycles of time: An extraordinary new view of the universe* (1st ed.). New York: Knopf.

Penzias, A. A., & Wilson, R. W. (1965). A measurement of excess antenna temperature at 4080 Mc/s. *Astrophysical Journal, 142*(07), 419–421.

Planck reveals an almost perfect Universe. (2013). [ONLINE]. http://www.esa.int/Our_Activities/Space_Science/Planck/Planck_reveals_an_almost_perfect_Universe. Accessed 13 May 2013.

Richard P. F. (2005). *Classic Feynman: All the adventures of a curious character,* First edn. New York: Norton.

Sagan, C. (1985). *Cosmos*. New York: Ballantine Books. Edition.

Sandage, A. (1997). *The universe at large. Key issues in astronomy and cosmology*. Cambridge: Cambridge University Press.

Sandage, A., & Perlmutter, J. M. (1990). The surface brightness test for the expansion of the μ, I – properties of Petrosian metric diameters. *The Astrophysical Journal, 350*, 481–491.

Vigier, J. P. (1988). *New ideas in astronomy*. Cambridge: Cambridge University Press.

Weinberg, S. (1972). *Gravitational and cosmology*. New York: Wiley.

2

Parallel Universes

> *Most sets of values would give rise to universes that,*
> *although they might be very beautiful, would contain no one*
> *able to wonder at that beauty.*
>
> – Stephen Hawking (1996)

Perhaps one of the most exciting – and unsettling – concepts in modern cosmology is the theory that we inhabit a parallel universe. Although this idea is widely used in the popular science media, the idea is not just in the realm of fiction any more. However, the extent to which it has proliferated often makes it seem like science fiction.

The proponents of parallel universes believe that there are vibrations of different universes everywhere. Probably, we are missing that subtle message, as we're just not in tune with the vibrations. There are possibly other parallel universes in our own living rooms. "The many worlds represent reality," wrote Michio Kaku, professor of theoretical physics at New York University and author of *Parallel Worlds: A Journey through Creation, Higher Dimensions, and the Future of the Cosmos* (2005).

Physicists hypothesize several levels of parallel universes. Some even envision an infinite number of parallel universes. Could there be people with your own memories and appearance? If we accept the idea of infinite universes, then there must be an outcome of every possible choice that we can't even imagine.

The scientific history of parallel universes begins with a doctoral thesis by Hugh M. Everett (UCIspace @ the Libraries 1973). Everett started his graduate work with John Archibald Wheeler at Princeton University in New Jersey. Applying the ideas of quantum mechanics, Everett made an outlandish conclusion even for today's standards. In his paper "The Theory of the Universal Wave Function," Everett argued that the universe is describable, in theory, by an objectively existing universal wave function. According to him a new universe is created every time we make an observation, and the wave function corresponding to each event does not collapse but gives rise to each independent reality.

S. Mathew, *Essays on the Frontiers of Modern Astrophysics and Cosmology*, Springer Praxis Books, DOI 10.1007/978-3-319-01887-4_2, © Springer International Publishing Switzerland 2014

Figure 2.1. Parallel universes. Are they a figment of the imagination or just unknown physical structures? (Image Credit: Wikimedia Commons)

This reality that he described is not the reality we customarily think of, but is a reality composed of many worlds.

The consequences Everett's assertions were staggering. And many physicists remain uneasy with it even today. To accept the notion that *everything that is possible can happen* is not an easy pill to swallow. Some researchers suggested that the way to achieve reconciliation is to drop the single universe view and to relate the multiplicity of frame representations of physics and mathematics to the many different physical universes viewed in physics (Benioff 2009). As we know, such opinions are still debated without any definite conclusion.

Though Everett's many worlds interpretation gained some popularity later on, for various reasons it fell apart. Additionally, Everett did not continue his work in theoretical physics, as he wasn't excited about working in academia, rather choosing a new career with military work. He died in 1982 at the age of 51.

PARALLEL UNIVERSES IN HINDU MYTHOLOGY

The concept of parallel universes may be novel and disconcerting to scientists, but it rests very comfortably within ancient Hindu cosmology. The *Puranas*,[1] Hindu religious texts thought to date back to between 500 and 1500 B.C., are replete with descriptions of many worlds whose inhabitants are ruled by kings in the human plane and Gods in a higher plane.

---

[1] The Puranas are a class of literary texts all written in Sanskrit verses.

The many Gods, in turn, belong to many different worlds and planes of existence. At the highest level of the hierarchy is the trinity, namely, Brahma, Vishnu,[2] and Shiva,[3] ruling the divine kingdoms. Brahma is the creator, who dreams the universe into being, which is maintained by Vishnu. What humans perceive as reality is in fact the dream of Brahma, misled by matter. Brahman[4] (not to be confused with Brahma) is the base of all reality and existence. Brahman is uncreated, external, infinite, and all-embracing. It is the ultimate cause and goal of all that exists. All beings emanate from Brahman; all beings return back to the same source. Brahman is in all things, and it is the true self (atman[5]) of all beings.

Well, the above mentioned themes coupled with myriad stories and an array of metaphors seem to be nothing more than the mythological stories widespread across the different cultures of the world. Yet, the metaphysical connotation of this idea and the cosmic formation in Hindu philosophy warrants some attention here, though this is not intended to infer any scientific conclusions.

In Hindu philosophy, the physical universe is a dream and it has only the kind of reality that a dream has. It is in a state of unceasing evolution, where names and forms arise and die out, but the true self remains unchanged. These material worlds float around while the cycles of creation and destruction continue endlessly. The ultimate reality is the Absolute (*brahman*), which transcends and includes everything and has been sought about extensively by some well-known physicists such as Erwin Schrodinger, Werner Heisenberg, and Niels Bohr. However, such interest in Oriental wisdom on the part of physicists has often been taken as an indication that the world view derived from physics is somehow deeply connected with that of eastern religions (Duquette 2011).

We have to maintain a clear distinction between mysticism and modern science. Exploring the subtlety of the universe or knowing the unknown seems to be the central theme of such writings. It should be clearly stated here that most physicists, even those who expressed an interest in Vedanta thoughts, consider that it is inappropriate to establish any solid conclusions by equating the insights from physics with mystical and religious ideas. Still, many physicists share a general view that ultimate reality cannot be known through direct experience such as personal or laboratory measurements. In fact, many physicists, while dealing with an entirely new world opened up by quantum theory and relativity, were often stranded in their effort to explain the experiments and observations.

In his book *Quantum Questions: Mystical Writings of the World's Great Physicists*, Ken Wilber (2001) remarked it is "the radical *failure* of [the "new"] physics, and not its supposed similarities to mysticism, that paradoxically led so many physicists to a mystical view of the world". Some even abandoned the pure scientific view to embrace the mysticism and philosophy, perhaps, as a last solace.

Among the many universes envisioned by physicists, one could exist in extra dimensions and might be physically very close to us. As creatures of a three-dimensional world,

---

[2] Vishnu is the second God in the Hindu triumvirate (or *Trimurti*). The triumvirate consists of three Gods who are responsible for the creation, upkeep and destruction of the world. The other two Gods are Brahma and Shiva.

[3] The role of Shiva is to destroy the universe in order to recreate it.

[4] In Hinduism Brahman is the ultimate reality, the power that makes the cosmos function.

[5] An individual soul or self. The ultimate goal in Hinduism is to achieve moksha through the realization that one's Atman and Brahman are the same thing.

we may not perceive it. But scientists hope that these universes might drop some clues to help us identify them, like ripples in a pond help us locate the actual disturbance. Modern scientific theories of creation and the world of particle physics help us develop a picture of our cosmos, which might be just one of many possible universes, provided the many-world interpretation will survive in coming years.

Since the mystical connotations have been of interest to many physicists, it's appropriate to discuss bit about the interest of Hindu Gods in Western culture.

WHY ARE HINDU GODS CELEBRATED IN THE WEST?

One of my fellow faculty members has a Ganesha deity on his table, which he had collected from India during a trip several years ago. Ganesha sits there displaying a deep sense of tranquility close to a computer in all his majesty and mystery in an American office setting. Another of my colleagues, who is an expert in world mythology, named her lovely daughter Kali, after the fearful and ferocious Hindu Goddess. In both cases, they ended up telling me that they simply liked the object and the name and gracefully skipped the religious and philosophical implications one could attach to these. But that made me wonder why Hindu Gods and Goddesses are getting so much attention in Western culture, though many of them are still unaware of it. Is it simply the curiosity that drives such affinity?

We all know that planets are named after Greek or Roman Gods. This is explicable given the Greek connection to early astronomy and the European inheritance of the ancient knowledge gained by Greeks. But often the names simply deceive. For example, the planet Venus is named after the Goddess of beauty, though now we know that the so-called beautiful planet's atmosphere is full of carbon dioxide with a floating mist of sulfuric acid that can corrode any flesh. What a strange beauty!!! Of course, beauty can be deceiving and dangerous. Obviously, ancient notions about planets and stars were skewed, though they laid the foundation of modern astronomy.

Sanskrit and Indian philosophy always had a broad appeal to philosophers and scientists. From Niles Bohr to Robert Oppenheimer, many scientists had a deep interest in the Vedas and Upanishads. Oppenheimer's remark from the *Bhagavad Gita,* "Now, I am become Death, the destroyer of worlds," has been widely written about; it came to his mind after Oppenheimer observed the first experimental detonation of an atomic bomb in the New Mexico desert.

Furthermore, the Hindu Gods easily blend into the human psyche. Many of them engaged in activities that we are used to. They, like us, loved, hated, killed and procreated. They enlightened their followers with words and deeds and explained material success and failure to the devotees, making it easier for them to overcome the delusion of both.

Another reason is the association Hindu Gods have with deeper philosophical nuance and its linkage to science. Theoretical physicist Fritjof Capra's *The Tao of Physics* is an international bestseller that explores and relates the depiction of the Nataraja posture with the continuous creation and destruction of particles and their different manifestations in the universe.

The strongest symbolic acknowledgment of this work is reflected in the Nataraja statue outside CERN, the European Organization for Nuclear Research, which has built the world's largest particle accelerator, the Large Hadron Collider (LHC), on the France-Switzerland border. The Indian government gifted a 6 ft Nataraja statue to CERN in 2004. It portrays Shiva's dance of creation and destruction, much like the dance of fundamental particles that generates and destroys matter and energy in the universe in various forms.

A plaque next to the Shiva statue captures the contemporary connotation of the metaphor of Shiva's cosmic dance from Capra's book: "Modern physics has shown that the rhythm of creation and destruction is not only manifest in the turn of the seasons and in the birth and death of all living creatures, but is also the very essence of inorganic matter and for the modern physicists. Then, Shiva's dance is the dance of subatomic matter. The metaphor of the cosmic dance thus unifies ancient mythology, religious art and modern physics."

It is ironic that, partly driven by capitalist market forces, the customer service representatives working for U. S. companies from their outsourced locations fake their name to better suit the English audiences, while many in the West find traditional Hindu names more and more attractive. There is little reason to resent this, as even Gods are subjected to the laws of nature for their survival.

Again, I learned that my colleague who is expecting soon will name her daughter Maya, the ultimate illusion. And, if anyone seeks a boy's name, a clear choice is Vishnu, another name that is vanishing from the Indian demographic landscape as virtual Bobs and Joes flourish and survive along with Pepsi and McDonald's. We cannot foresee how future generations will perceive the world. They might not be interested in any illusion, as in years from now they may have transformed into beings without any name and desire to know anything. Then the great Maya will dissolve in Brahman.

OBSERVATIONAL EVIDENCE

The only observational evidence, purportedly, in support of parallel universes has come from the aforementioned NASA's Wilkinson Microwave Anisotropy Probe (WMAP). While analyzing the cosmic microwave background (CMB) radiation data, scientists discovered evidence of a huge void spanning almost 1 billion light-years (1 light-year[6] is approximately 10 trillion km). The void in the infant universe represents the absence of any material, which otherwise should have become stars and planets. None of the current cosmological theories can explain such huge voids in the data. Some physicists interpret this as the unmistakable imprint of another universe beyond the edge of our own. Physicist Mersini-Houghton proposed a model of entangled universes, under which they predict two huge voids, not just one (Frankel 2011). One of them has been found by WMAP data, and new data is expected to reveal a second similar void. The recently launched Planck satellite by the European Space Agency, whose exceedingly sensitive detectors measure CMB radiation and which captured its first image recently, may be able to ascertain this second void.

Even if everyone agrees these features are caused by shadow universes, we still could not deduce anything about them aside from their ghostly thumbprint.

---

[6] A light-year is a unit of distance. It is the distance that light can travel in one year. Light moves at a velocity of about 300,000 kilometers (km) each second. So in one year, it can travel about 10 trillion km. More precisely, one light-year is equal to 9,500,000,000,000 km.

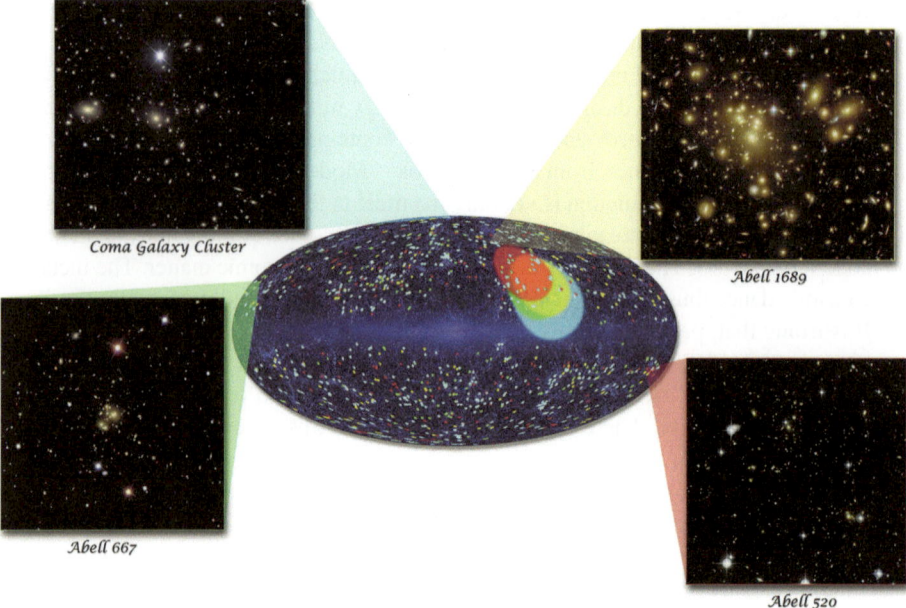

Figure 2.2. The effect of multiverse. The colored dots are clusters within one of four distance ranges, with *redder colors* indicating greater distance. Colored ellipses show the direction of bulk motion for the clusters of the corresponding color. Images of representative galaxy clusters in each distance slice are also shown (Image credit: NASA/Goddard/A. Kashlinsky et al.).

COSMIC PARALLELS?

Whether Vedic cosmology, as some suggest, has scores of other eerie parallels with some of the most cutting-edge recent cosmological theories of multiverse, oscillating universe, and the Big Bang, so it should come as no surprise that some of the greatest minds in science have turned to these 3,500-year-old cosmological ideas for inspiration and questions about our enigmatic universe. Some physicists have shown deep interest in philosophical aspects of Upanishads.[7]

Striking as these similarities are, it does not mean that modern science has vindicated Vedic cosmology. For one, these modern cosmological theories may themselves be disproved and, besides, many physicists remain skeptical of these theories. Some cosmologists, for instance, question the whole notion of multiverse. According to them, if the universe encompasses everything we know or ever want to know, it rules out any room for parallel universes. Some other models suggest that universes are finite in number and restricted by mathematical formulations.

---

[7] The latest of the writings to be considered part of the Vedic period, written between the eighth and third centuries BCE. These are collections of stories, discussions, and instructions addressing issues of the relationship between the human and the ultimate realms.

Even so, the idea that all structures that exist mathematically also exist physically is the foundation of the parallel universe concept. This hypothesis, known as the ultimate ensemble, predicts the existence of all universes that can be defined by mathematical equations. But many physicists disagree on the grounds that not all mathematical structures are well defined.

Nevertheless, the concept of many worlds or parallel universes, which would have invited the ridicule of mainstream physicists, as it did when it was first proposed more than half a century ago (1957) by American physicist Hugh Everett, is currently one of the hottest trends in theoretical physics. The multiverse theory is the inevitable result of quantum mechanics, which represents a set of multiple probable states for a particle. When an observation is made, the particle chooses one of the multiple states measured by the observer, and the other states collapse. This is the most basic principle behind the existence of many universes, or multiverses.

Quantum physics, the study of the minutest particles that make up matter, has been remarkably successful, but it reveals a picture of a quantum reality so strange that our minds are unable to grasp it. For instance, quantum mechanical experiments have proved that objective reality is unlikely to have a separate existence. The nature of the ultimate reality is intertwined with our actions, and uncertainty rules when we observe it. This is contrary to our common sense view that reality has an existence independent of the observer. In the realm of the smallest particles, however, objective reality is not an absolute entity that can definitely be measured, as is true in classical physics.

Erwin Schrodinger, a leading theorist in quantum mechanics, who had a lifelong interest in philosophy, wrote (2012): "From all we have learnt about the structure of living matter, we must be prepared to find it working in a manner that cannot be reduced to the ordinary laws of physics. And that not on the ground that there is any 'new force' or what not, directing the behavior of the single atoms within a living organism, but because the construction is different from anything we have yet tested in the physical laboratory."

This is not very far removed from the concept of Brahman – the self-existent, immanent, and transcendent supreme and ultimate reality. Brahman is the fundamental divine cause of everything in and beyond the universe. The nature of Brahman is explained as personal and impersonal, and it is the source of the creation of the universe and Gods. The universe and all the objects in it are the manifestation of a fundamental reality, which we don't know yet. We understand the universe as an exchange of matter and energy, and scientists often disagree about the fundamental building blocks of these phenomena.

The absence of objective reality implies that the material world could be an illusion – or, what Hindu sages called Maya. In Hinduism, Maya[8] is the natural illusion that the material world is the only reality. It is a skewed perception, albeit commonly held, to believe that the material world is the fundamental reality. Maya, which has its roots in the Upanishads, denotes the power of God to make human beings believe in an illusion. The material world is the manifestation of Brahman, the infinite and immortal reality that is responsible for matter, energy, space, time, and every being. The scriptures and philosophies seek to unveil the illusion to learn the ultimate truth. However, such descriptions are

---

[8] Maya is the illusion or the skewed perspective with which we experience the universe.

purely philosophical in nature, definitely interesting, and our rational thought demands a transparent and logical procedure to know the ultimate truth.

Science employs its own methods and procedures toward that goal. Our universe and its subjects are creatures of spirit coated with matter that conceals the spirit from the light. Science can demonstrate its processes, is based on rationality and logic, and can be comprehended, but that is somewhat lost in the scheme of the ancient wisdom. Consider, for example, this Upanishad expression, "That which permeates all, which nothing transcends and which, like the universal space around us, fills everything completely from within and without, that Supreme non-dual Brahman –That Thou Art."

Our interpretations of the evolution of the cosmos, once based on myths and legends, have advanced markedly in the last century. We live in a universe that is beaming with billions of galaxies that are continuously expanding. We now know that this expansion is accelerating, and we will never know what is beyond our cosmic horizon, as no light will ever reach us from the expanding universe. It is ironic that we need many universes for the existence of our own universe!

The scientific approach always assumes that fundamental reality is different from us, and we are independent observers seeking truth. But many researchers now believe that we must rethink this assumption. Our observation impacts the observed reality, because we are part of it. This essentially is the Advaita[9] philosophy, a cornerstone of Hinduism, which asserts that Brahman (ultimate reality) and Atman (self) are the same. Such a non-dualistic approach is advocated by many modern researchers who argue that the effect of observation either changes reality or creates new realities.

Human eyes operate using the visible light of the electromagnetic spectrum, allowing us to view only objects that emit light, which comprise only a small fraction of the universe. In modern times, telescopes augmented the unaided eye in the hunt for the unknown. Operating from ground and space, these telescopes scan the cosmos to draw pictures of material objects. Even with telescopes, we exploit the electromagnetic radiation to weave images of the cosmos. Whether it is gamma rays, X-rays, microwaves, or visible light, throughout human history we have been dependent upon different forms of light to learn of cosmic events that document our own history.

Here, let us discuss the most popular telescope of our time – the Hubble Telescope, known as the mirror on the universe.

THE MIRROR ON THE UNIVERSE, NOW TWENTY-TWO YEARS OLD

While supporting Galileo in the backdrop of his conflict with the Catholic Church, a cardinal once remarked, "Bible teaches how to go to heaven, not how the heavens go." Now, 400 years after Galileo attempted to know the heavens using his "spy glasses," our machines can narrate the story of the heavens in much more detail.

The heavens narrate their stories in a distinct manner and allow humans to discern the mystery of creation and evolution. Up to now, humans could accomplish this only through

---

[9] Advaita means non-dual and is a prominent school of thought in Hindu philosophy. This doctrine identifies the self (atman) with the ultimate reality (Brahman) and negates any real distinction between the individual and the entire universe.

the decoding of light. This is because light is a messenger that can convey the untold chronicles of the cosmos, which has been a great source of myths and legends ever since the beginning of humankind. For the ancients, the heavens were the citadel of Gods who visited them for various reasons and often punished them with fiery objects. The naked eye had been the only means to investigate the elements of the cosmos, and it changed forever in 1609. Galileo, the father of modern astronomy, developed a new scientific world when he used the power of the telescope to explore the heavens. He narrated the accounts of his observation in *The Starry Messenger* published in 1610.

Telescopes are often referred to as time machines, as they escort us back in time. When we peep at a star or any other object a few million light years away, we are in fact seeing that object as it existed a few million years ago. Since Galileo's first use of the telescope, scientists have been improving the power of telescopes to gaze at the unfathomable universe. Now, 400 years after the Galilean adventure, modern astronomers are on the verge of investigating the frontiers of the known universe. A variety of telescopes, operating from ground and space, aid them in this process. If the Galilean 'spy glasses' were able to reach just the backyard of our galactic neighborhood, the modern era telescopes take us closer to the moments of creation known as the Big Bang.

Among these machines is the world's most famous telescope, the Hubble Space Telescope, which turned 22 on April 24, 2012. During the last two decades of operation, it saw the birth and death of stars and captured many turbulent cosmic collisions. It granted us an exotic vision to enjoy the wonders that lie in the tempestuous cosmic ocean. Some call it "The Mirror on the Universe," while others describe it as "The Eye on the Sky." It continues to beam hundreds of images back to Earth every week.

Figure 2.3. There are many descriptions for the Hubble telescope, including the eye on the sky or the mirror on the universe (Image credit: NASA).

As mentioned in Chapter 1, named after the American astronomer Edwin Hubble, whose observations in the 1920s supported the theory of an expanding universe, Hubble has been in continuous action for the last two decades. Since its launch in 1990, most of its original instruments have been upgraded or replaced by service missions. Hubble, located at about 565 km above Earth's surface, approximately the size of a school bus, completes one full orbit around Earth in 97 min.

In addition to many startling discoveries, the Hubble images have become the artwork of the cosmos. The Hubble Ultra Deep Field (HUDF), completed in 2004, is an image of a small region of space created using Hubble data accumulated over a period of 4 months. In fact, some of these objects date back to the baby universe, approximately 13 billion years ago, when the galaxies were just forming from the seeds of the Big Bang. This particular image contains an estimated 10,000 galaxies in different shapes and sizes. Each of them might contain billions of stars and many possible planetary systems. Scientists were perplexed at the mere existence of such a large number of galaxies, and some even dubbed them as "Kingdoms of Heaven."

The mystery surrounding the creation and existence of the universe reaches out to us in the form of light energy. Hubble has done more than any other modern telescope to garner that energy and to paint a picture of the history of the universe for coming generations. Edwin Hubble observed and measured the departure of galaxies using a technique known as the redshift in physics. Now we know that the galaxies not only depart from each other, but their exodus is accelerated by the inexplicable dark energy.

The latest and last Hubble repair mission was conducted in May 2009, extending the life span of the telescope for another 5 years. The instruments on the telescope can observe the edges of the universe in visible light, ultraviolet and infrared ranges of the electromagnetic spectrum. HST is located at about 565 km above Earth's surface with an approximate size of a school bus. The website devoted to HST (http://hubblesite.org) provides every detail and discoveries of the telescope, and enable the public to track every moment of its voyage.

If our current notion of the universe is true, in the far future our own Milky Way Galaxy will be left alone in the galactic playground, with other galaxies having receded to unknown parts of the cosmos. The finite speed of light will not overcome the unlimited space that would be created among the galaxies due to the accelerating nature of their retreat. This could lead future generations to assume that their galaxy is the same as the universe. If preserved, the Hubble images will enlighten our descendants with the chronicle of that ultimate isolation.

HOW FAR BACK CAN WE SEE?

We believe what we see, but astronomy has long taught us that our eyes deceive. What we see today might be the drama of the cosmos unfolded long ago. Past, present, and future lose their meaning in the vastness of the universe.

Recently, researchers and telescopes detected the most distant object in the visible universe. While we celebrate the discovery, the fact is that this particular object likely no

Figure 2.4. As far as NASA's Hubble space telescope can see. This view of nearly 10,000 galaxies is the deepest visible-light image of the cosmos. Called the Hubble Ultra Deep Field, this galaxy-studded view represents a "deep" core sample of the universe, cutting across billions of light-years (Courtesy of NASAimages.org-Galaxies, Galaxies everywhere).

longer exists. So, even as we wonder how far we can see to the edge of the universe, it also invites the fundamental question: Can we trust what we see?

Four hundred years after Galileo peeped into the heavens using the first telescope, human civilization set its sights on the edge of the visible universe. The gamma ray burst (GRB) named GRB 090423 (Tanvir et al. 2009) is the most remote object we have ever seen in the cosmos. There are no magic machines in the foreseeable future to lead us further.

This GRB, estimated to be 13.1 billion light-years away, is one of the brightest stellar explosions recorded. NASA's space telescope SWIFT spotted the event first in April 2009 and then scores of ground-based telescopes took over.

Astronomers use redshift as a tool for determining distances in the universe. The redshift is the wavelength or frequency shift of light as it travels. This is similar to the changes in the frequency of an ambulance siren as it passes by. We perceive the declining frequency or lower pitch of sound as it travels, even though the ambulance is producing it at the same frequency. Analysis of light from cosmic events, such as a GRB enables researchers to measure the distance and origin of such events. Light is a messenger that carries the details of its journey in its wavelength spectrum. It also means that if light does not reach us, there is no way of knowing what is out there in the farthest corners of the universe.

GRB 090423 is the aftermath of the explosion from one of the early stars in the universe. Since this event occurred 13.1 billion years ago and the universe is considered to be 13.7 billion years old, this first-generation star had a life span of about 630 million years – a young star by cosmic standards. It is usual for massive stars to end their life at younger ages compared to less dense stars, which survive billions of years.

When mid-sized stars, such as our Sun, finish their main sequence life, they end up as "white dwarfs," a relatively quiet event. However, massive stars send out the message of their demise as waves after a violent death in the form of supernovae. During this process, the core of the star transforms into a highly dense neutron star, sending the outer layer of stellar masses to form a nebula.

A nebula serves as the feeding grounds for the next generation of stars. Nebulae teach us that death and birth are cyclic in the universe rather than absolute transformations. Our Solar System was once a part of a nebula created from the death of a star caused by a supernova explosion. In nature, creation-preservation-destruction are continuous cycles, without beginning or end.

A supernova explosion (more details about supernova will be discussed in Chap. 6) in the past created a nebula that in turn became the birthplace of our star, the Sun, and its planets. Scientists believe our Sun is a third-generation star. Millions of years later, organic molecules formed in one of the planets and evolved to become intelligent beings – humans. In essence, all the heavier elements, such as carbon, oxygen, and nitrogen that make up our body came from supernova explosions. The elements heavier than hydrogen were created in the interiors of stars and then expelled into space, to be integrated into later stars. As the astronomer Carl Sagan once noted, we are made of "star stuff."

If we want to see our cosmic ancestry, nebulae are the best to place to look at. The Eagle Nebula, which is 7,000 light-years from Earth, is one of the most admired nebulae, thanks to the Hubble Telescope, which captured this nebula in all its majesty. The Eagle Nebula shows huge columns of gas and dust, light-years across, known as the Pillars of Creation. Please remember that the final illustration is not the actual color, but is closer to

Figure 2.5. Gamma ray bursts are so intense despite happening halfway across the universe, they sometimes can be seen briefly with the unaided eye (Image credit: ESO).

what we would see if we could see all the wavelengths of color. The picture is a combination of different images taken through different filters, and has been processed to eliminate cosmic radiation and other distractions. The pictures show that new stars are born in this stellar nursery. Astronomical calculations reveal that these pillars vanished 6,000 years ago due to a nearby supernova explosion. Yet, we will continue to see the intact pillars for another 1,000 years as the message of that destruction in the form of light that has yet to reach us. The pillars are truly an impression of the past, and they tell us that time is an illusion in which we live, along with space, and is created by the movement of objects and perceptions. In some sense, time exists because we are bound to things through our senses.

The stellar nursery, aptly named, pillars of creation with its majestic appearance teaches us not only about our cosmic history, it reveals the grant illusion unfolding in huge scales in front of our own eyes. The gravitational forces churn the cosmic material to prepare for celestial births. It is not hard to imagine that a few billion years ago, our own star, sun had a similar origin in the shadows of a nebula we could call solar nebula.

Supernova explosions are illustrious events among astrophysicists because they provide a rare opportunity for researchers to study distant parts of the cosmos. Even our most sophisticated telescopes are unable to spot a star at those distances, but a supernova outshines even its mother galaxy, providing an exceptional chance to look at it.

The stellar explosions more powerful than supernovas are known as hypernovae. These are events capable of sterilizing life in their cosmic neighborhoods. In a galaxy such as our own, the Milky Way, these events occur only rarely – once every 100,000 years. Given the billions of galaxies spanning the universe, however, it is a common event in the universe. If the supernova explosion leads to neutron stars and black holes, scientists think that hypernovae might create something more than just black holes. In the case of hypernova, the core of the star turns into a black hole, but the outer material falls back into the core, resulting in gamma-ray bursts (GRB). The GRBs are said to be the "birth cry" of black holes from the farthest corners of the universe.

When we look at the Sun, in fact we are seeing the Sun as it existed 8 min ago – the time light takes (at a speed of 186,000 miles per second) to travel the 93 million miles between the Sun and Earth. Even if the Sun disappears in a cosmic event, we will continue to see the non-existent Sun for 8 min! We perceive the past of the Sun as our present Sun. By the same reasoning, what we are seeing of the GRB is an event that took place 13.1 billion years ago. This GRB and the resulting nebula must be long gone, but earthlings see it now as the "messenger light" from that occurrence, as it has just reached us in this corner of the Milky Way Galaxy. We are looking at this GRB today and seeing yesterday!

Under the Big Bang theory, cosmologists believe the universe has been expanding since its birth 13.7 billion years ago. When the waves originating from this hypernovae began their journey, the universe was much smaller than its current size. As the waves propagated throughout the universe, they had to travel greater distances as the departing galaxies created new space between them. The gamma rays encountered this new space as extra distance on their voyage. Thus, the successor of this event, if it exists, must be much farther than 13.1 billion light-years. Probably, we will never know about it as its distance to our galaxy has increased so much that the light or any electromagnetic radiation it emits will never reach us.

The visible universe represents the space we can conceive in the sense that radiation from there will reach us. There must be much more to explore beyond the visible universe, but we have no means of knowing. As the universe continues its expansion, eventually our Milky Way Galaxy will be left alone in the universe with all the others departed to farther reaches. In the absence of any information from other galaxies, the mortals on this galaxy might assume that their universe is their own galaxy.

So how far can we see? The puzzling answer is around 13 billion light-years. Our best telescopes can see a few million years after the origin of the universe. But we cannot see anything before the point when light did not emerge out of the baby universe. Asking the question, "How far can we see?" is actually the same as inquiring, "How far back in time can we see?" The expanding universe imposes a limit on our view, and so we will not see anything beyond 13 billion light-years ago. And when we see it, unfortunately it will not be there anymore!

The early universe was not transparent to light, which implies that we can write the history of our cosmos only up to the point when the first light began its journey, long after the creation of the universe. Additionally, the interaction of light with matter distorts the details of the information it carries. However, unlike light, gravitational waves propagate through the cosmos without reflection or refraction. That could potentially allow us to create a purer picture of the cosmos beyond the levels light allows us.

Nevertheless, it would be an incredible proposition to gather the evidence for parallel universes. We haven't been able to know the edges of our own universe, and it may not be possible to do so as the expanding universe rewrites its own boundaries. Yet, the concept of multiverses would remain as powerful as our own universe. Carl Sagan once remarked (1980), "The universe is not required to be in perfect harmony with human ambition. The universe seems neither benign nor hostile, merely indifferent."

As the creatures of this universe, our dreams and imaginations will pursue not just the copies of our planet but also the replicas of our own universe.

Figure 2.6. The pillars of creation. Stars are born (Image credit: NASA).

REFERENCES

Ananthaswamy, A. (2007). *Mystery of a giant void in space.* http://www.newscientist.com/article/ mg19626355.300. Accessed 12 July 2011.

Ashtekar, A. (1999). *Quantum mechanics of geometry.* http://arxiv.org/abs/gr-qc/9901023. Accessed July 2011

BBC – Religions – Hinduism: Vishnu. (2009). BBC – Homepage. http://www.bbc.co.uk/religion/ religions/hinduism/deities/vishnu.shtml. Accessed 7 Oct 2010.

Benioff, P. (2009). A possible approach to inclusion of space and time in frame fields of quantum representations of real and complex numbers. *Advances in Mathematical Physics, 2009*, article id 452738, 22 pages, 2009. doi:10.1155/2009/452738.

Bohm, D. (1951). *Quantum theory.* New York: Prentice-Hall.

Davies, P. C. W. (2004). Multiverse cosmological models. *Modern Physics Letters A, 19*(10), 727–743.

Davies, P. (2006). *The goldilocks enigma.* New York: Mifflin.

Dorman, E. R. (2011). Hinduism and science: The state of the south Asian science and religion discourse. *Zygon: Journal of Religion & Science, 46*(3), 593–619. Academic Search Premier, EBSCOhost, viewed June 2, 2012.

Duquette, J. (2011). Quantum physics and vedanta': A perspective from bernard d'espagnat's scientific realism. *Zygon: Journal of Religion & Science, 46*(3), 620–638. Academic Search Premier, EBSCOhost, viewed May 19, 2013.

Everett III, H. (1957). "Relative state" formulation of quantum mechanics. *Reviews of Modern Physics, 29*, 454–462. http://www.scientificamerican.com/article.cfm?id=hugh-everett-biography

Frankel, M. (2011). Time and the multiverse | plus.maths.org. 2012. Time and the multiverse | plus. maths.org. [ONLINE] Available at http://plus.maths.org/content/time-and-multiverse. Accessed 14 Mar 2012.

Galileo and Theology (Cosmology: Ideas). (2013). Galileo and theology (cosmology: ideas). [ONLINE]. http://www.aip.org/history/cosmology/ideas/galileo.htm. Accessed 14 Mar 2013.

Gross, D., Henneaux, M., & Sevrin, A. (Eds.). (2005). *The quantum structure of space and time: proceedings of the 23rd Solvay conference, Brussels, Belgium, December 2005*. New Jersey: World Scientific Press.

Hawking, S. W. (1996). The Illustrated Brief History of Time (Updated and expanded edition, updsted sub edition). New York: Bantam.

Horowitz, G. T. (2005). Spacetime in string theory. *New Journal of Physics, 7*, article 201.

Hugh, E. (1957). *Reviews of Modern Physics 29*, 454–462. Bibcode 1957RvMP...29..454E.

Jack Ng, Y. (2007). Holographic foam, dark energy and infinite statistics. *Physics Letters B, 657* (1–3), 10–14.

Kaku, M. (2005). *Parallel worlds: A journey through creation, higher dimensions, and the future of the cosmos*. New York: Anchor Books.

Kaku, M. (2008). *Physics of the impossible: A scientific exploration into the world of phasers, force fields, teleportation, and time travel*. New York: Doubleday.

Madore, J. *Noncommutative geometry for pedestrians*. http://arxiv.org/abs/gr-qc/9906059

Raffa, F. A., & Rasetti, M. (2009). Natural numbers and quantum states in fock space. *International Journal of Quantum Information, 7*(Supplement), 221–228.

Sagan, C. (1980). *Cosmos*. New York: Random House.

Sagan, C. (1994). *Pale blue dot: A vision of the human future in space*. New York: Random House.

Sagan, C. (2000). *Carl Sagan's cosmic connection: An extraterrestrial perspective*. Cambridge: Cambridge University Press.

Schrodinger, E. (2012). *What is life? With mind and matter and autobiographical sketches* (Canto classics Reprintth ed.). Cambridge/New York: Cambridge University Press.

Tanvir, N. R., et al. (2009). A gamma-ray burst at a redshift of z=8.2. *Nature, 461*(7268), 1254–1257.

Tegmark, M. (2003). Parallel universes. *Scientific American*. April 14, *282*, 41–51.

Tegmark, M. (2008). The mathematical universe. *Foundations of Physics, 38*(2), 101–150.

UCIspace @ the Libraries – "The Theory of the Universal Wave Function," long thesis as published, 1973. 2013. UCIspace @ the Libraries – "The Theory of the Universal Wave Function," long thesis as published, 1973. Available at http://ucispace.lib.uci.edu/handle/10575/1302. Accessed 10 July 2011.

Wheeler, J. A. (1957). Assessment of Everett's 'Relative State' formulation of quantum theory. *Reviews of Modern Physics, 29*(3), 463–465.

Wilber, K. (2001). *Quantum questions: Mystical writings of the world's great physicists* (Revisedth ed.). Boston: Shambhala.

Wolfram, S. (2002). *A new kind of science*. Champaign: Wolfram Media.

Zukav, G. (1979). *The dancing Wu Li masters: An overview of the new physics*. New York: Morrow and Co., 1979,Many worlds theory. pp. 106–110, 319, 321. Described by [162].

3

Once Upon a Time, There Was No Time or Space

Time is of your own making;
Its clock ticks in your head.
The moment you stop thought
Time too stops dead.

–Angelus Silesius, sixth-century philosopher and mystic
poet (*The Book of Angelus Silesius,* 1985)

Salvador Dali's painting "The Persistence of Memory," is a splendid illustration of a deep mystery – the concept we call time. This surreal painting shows ants devouring a clock and time melting into oblivion. Is time an illusion that melts away as depicted in the painting or a fundamental reality that weaves the fabric of the cosmos with space?

Perhaps the mystery of time holds the key to the ultimate truth that humanity is seeking. It remains to be seen whether time will persist forever or fade away, as in Dali's paintings. In the fundamental realm, time may be the result of some missing knowledge about reality rather than a feature of reality.

Throughout history, humans have been intrigued by the nature of time. Philosophers such as Plato, who contemplated time as immortal, and Newton, who elevated time as eternal, might not accept the death of time. However, the last century witnessed the descent of time from immortal to mortal through Albert Einstein's equations. Now, the descendants of Einstein are trying to annihilate time, which had been wounded in the revolution unleashed by the theory of relativity.

S. Mathew, *Essays on the Frontiers of Modern Astrophysics and Cosmology*, Springer Praxis Books, DOI 10.1007/978-3-319-01887-4_3, © Springer International Publishing Switzerland 2014

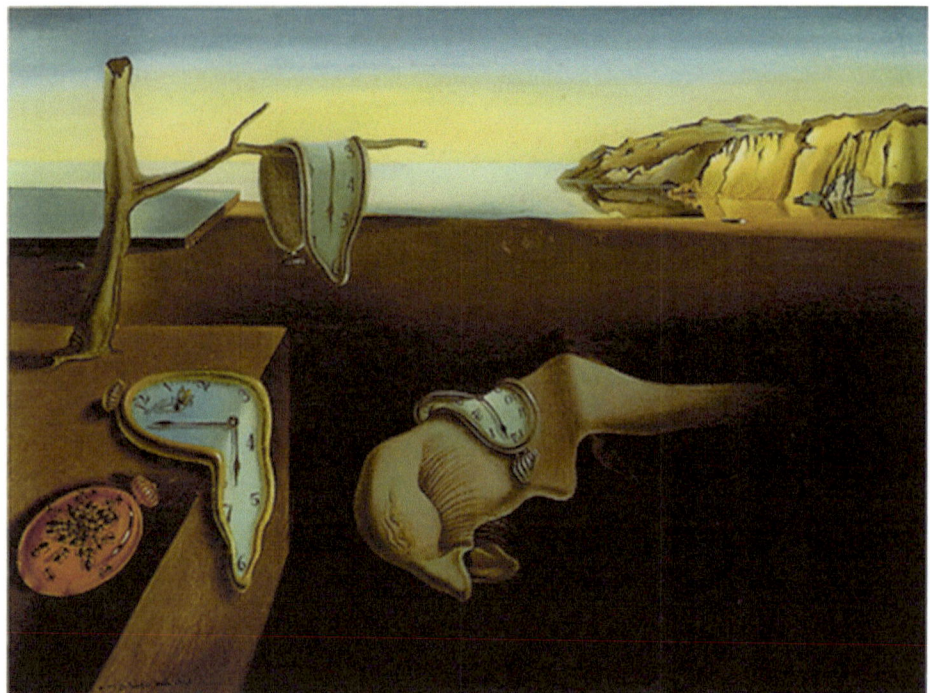

Figure 3.1. The persistence of memory-time is melting away. Real or surreal? (Creative Common Image courtesy of Stephanie on Flickr. Used with permission.)

DEATH OF TIME

We all have heard that time and tide waits for no man. But what if there is no time as we like to think of it? Moreover, can our modern physics survive without the need for time?

In an essay written about the nature of time theoretical physicist and cosmologist Carlo Rovelli explained why we must forget time (2008). The framework that he developed advocates timeless universe. He pointed out: "We never really see time. We say we measure time with clocks, but we see only the hands of the clocks, not time itself. And the hands of a clock are a physical variable like any other. So in a sense we cheat, because what we really observe are physical variables as a function of other physical variables, but we represent that as if everything is evolving in time." (Discover magazine 2007)

Some theoretical physicists argue that information about the measurement of time could be replaced with correlated observations in space. As an example, in 2008, *New Scientist* magazine reported as Rovelli saying, "I can tell you that I drove from Boston to Los Angeles, but I passed first through Chicago and later through Denver. Here I am specifying things in time. But I could also tell you that I drove from Boston to LA along the road marked in this map."

In other words, the dynamics of the universe is not progressing as evolution in time – the currently held notion in science – but as a network of correlated variables. Time emerges as an effect of the actions of correlated variables. This is similar to saying that temperature results from the interaction of a large number of molecules. Similarly, time is the manifestation of the interaction of some variables, rather than a fundamental quality of the universe.

In classical physics, time is a river flowing simply in the forward direction. The observer or external forces like gravity have no impact on the flow of time. Everything we know surfaced, survived, and departed along the course of this river, such as men and women, science and music, thoughts and feelings. Nothing could endure but time. Like an arrow shot by the Big Bang, time traveled (existed) for about 13.7 billion years along with its anomalous twin, "space."

Einstein's laws of relativity (1920) rebuked an absolute status for time and declared that time is as relative as space is, and together they make the fabric of the cosmos. This space-time geometry of the universe is one of the pillars on which the whole of modern cosmology is founded.

The general theory of relativity demonstrated that clocks in a stronger gravitational field tick at a slower rate, while the special theory of relativity argued that clocks moving at high speeds will appear to tick slower than non-moving ones.

As a modern-day example, atomic clocks on board GPS satellites move faster by about 45,900 ns (a nanosecond is 1-billionth of a second) a day because they are in a weaker gravitational field, as they are at a higher altitude than atomic clocks on Earth's surface. Similarly, GPS clocks run slower by about 7,200 ns a day, as predicted by the special theory of relativity, owing to the fact that the satellites have orbital speeds of about 3.9 km per second in a frame centered on Earth. To offset these time variations, the satel-lite clocks are reset before launch to compensate for these predicted effects. This is the consequence of Einstein's theory, something that was unimaginable even for the scientific community when Einstein predicted it decades ago.

Even the time span of our present day will change in a strong gravitational field. The current 24-h day could have extra time added to it if the gravitational effect on this planet was different than what it is now, or it could have less. In fact, some calculations show that in a few 100 million years, Earth's rotation will slow down, making a day 25 h long!

If time is a dimension, like space, as suggested by the theory of relativity, it must be plausible to go back and forth in time as we do in other dimensions, such as up and down, left and right, and forward and backward. Yet, in our universe, the cosmological arrow of time has not been known to reverse direction since the beginning of the universe.

Contrary to our expectations, however, the laws of physics do not discriminate between past, present, and future. We are obsessed with time and timekeeping, and it follows us like a shadow all the time. Physics tells us that all moments exist equally, at once. It's only our perception that distinguishes the present from the past or future. Some cosmologists suggest that all the frames that belong to future, past, and present exist all together. It's like a movie: the physical reality that unfolds every day in front of us is already there.

Though back (past) and forth (future) time travel is theoretically permitted by the laws of physics, it is riddled with paradoxes. Among the most well-known is the "twin para-dox," in which one twin travels at relativistic speeds (comparable to that of light) through the universe for many years, leaving his twin brother behind on earth. When he returns, he

Figure 3.2. We measure the time using clocks; however, that does not explain what time is on a fundamental level. The clocks, both mechanical and biological, pulse differently on different frames of references – the illusion of time.

would have aged much less than his sibling. The theory of relativity dictates that in a fast-moving frame of reference, all clocks, and thus aging, would move slowly.

Another paradox offers an even grimmer outcome. What if you travel back in time and murder your grandfather? Some physicists suggest the laws of physics must always conspire to prevent travel into the past and thus the impossibility of such a paradox. However, a few others believe that such actions simply cause space-time to branch off into a new parallel universe that doesn't interfere with the current one.

Modern science is not the first to attempt to comprehend the illusive nature of time. Mythology likewise sought such answers, too. For instance, in Purana,[1] Narada asks Vishnu about the meaning of Maya, to which Vishnu replies that it is the illusion generated as a consequence of our interaction with the physical world.

This view is echoed in Buddhism. "From the Buddhist perspective, time is the experience of being present right now, in this very moment. We in the West, however, like to measure things. In this way, clock time gives us a sense of coherence and stability. But in

---

[1] Puranas constitute the sacred literature of Hinduism along with Upanishads and Vedas.

terms of our inner lives, no time exists except for what is happening in the present moment," explains Joan Halifax, a Zen Buddhist and a distinguished scholar at the Library of Congress.

The subjective nature of time is implicit in human life, not just in cosmology or physics. Some events speed up time for us, while others make it crawl. "Put your hand on a hot stove for a minute, and it seems like an hour. Sit with a pretty girl for an hour, and it seems like a minute. That's relativity," Einstein once explained.

Neuroscientists, examining how the brain perceives time, say that the sensation of the passage of time depends on how richly the memories are laid down. Tragic events are likely to be perceived as more recent than they really are or than events that are not registered as strongly.

Many researchers believe that information and the mode of processing provide the sense of past and future. It is the "flow of information" rather than the "flow of time" that creates this impression – much like a system stores past information (memory) to be recollected but focuses attention on the most recent information.

Often, biologists refer to a biological clock based on which our cellular actions are carried out. So if we are completely insulated from our surroundings, the biological clock might function to generate the perception of time passing by.

Not everyone agrees. "Time has little impact on biology," says (as reported in *Forbes* magazine, 2008). Michael West, a gerontologist who teaches at the University of California, Berkeley. "From a gerontologist's standpoint, biological time is not wear-and-tear, it's a genetic program," argues West. "It's sort of like a time bomb. The cells are programmed to last just long enough for us to bear children, and no longer." The absence of clocks in our system does not guarantee an eternal life; we will fall apart anyway.

In elementary particle physics, particles can travel back and forth in time. This is one way of explaining particle-antiparticle annihilation. (The particle-antiparticle situation will be discussed more fully in Chapter 4.) However this is one way of interpreting the mutual annihilation of particle-antiparticle pairs. There is definitely a different way of explaining such results without considering the backward movement in time.

Why do we have such contradictions when we deal with time? Why can't we treat time in a coherent manner? The answer is, as embarrassing as it might be, that physics has not fully solved the mystery of time. Our collective thoughts and wisdom haven't experienced the reality to its full depth.

There is reason to be skeptical about the methods of science, at least in the case of time, as English mathematician and philosopher Alfred North Whitehead wrote, "Science can find no individual enjoyment in nature: Science can find no aim in nature: Science can find no creativity in nature; it finds mere rules of succession. These negations are true of natural science. They are inherent in its methodology. The reason for this blindness of physical science lies in the fact that such science only deals with half the evidence provided by human experience" (1968). At the same time, we need to find solace in the fact that science has far to go and will continue beyond the constraints of time.

Let me emphasize that the death of time is not a done deal. But this whole theme teaches us that in our search for the whole picture there is something missing. That missing link is related to the concept of time. Whether time will survive or perish which remains to be seen. Maybe we have to re-invent the meaning of time.

The orthodoxy of science sometimes hinders the thought process. This is demonstrated by an anecdotal story that is attributed to Neils Bohr, the father of atom. A physics student at the University of Copenhagen was once asked the following question in an exam: "Describe how to determine the height of a skyscraper using a barometer."

The student answered: "Tie a long piece of string to the barometer, lower it from the roof of the skyscraper to the ground. The length of the string plus the length of the barometer will equal the height of the building."

This answer, as one can imagine, annoyed the examiner and the student was given a failing grade on the exam.

Unsatisfied with the decision, the student appealed and the university appointed an independent arbiter to decide. The arbiter judged that the answer was indeed correct, but that it did not display any noticeable knowledge of physics.

To resolve the problem, it was decided to call the student and allow 6 min for him to provide an oral answer. For a few minutes the student sat in silence, his forehead creased in thought. When the arbiter pointed out that time was running out, the student replied that he had several extremely relevant answers but could not decide which to use. "Firstly, you could take a barometer up to the roof of the skyscraper, drop it over the edge and measure the time it takes to reach the ground, but too bad for the barometer. If the sun is shining you could measure the height of the barometer, then set it on end and measure the length of its shadow. Then you measure the length of the skyscraper's shadow, and thereafter it is a simple matter of proportional arithmetic. If you wanted to be highly scientific, you could tie a short piece of string to the barometer and swing it as a pendulum, first at ground level, then on the roof of the skyscraper. The height of the building can be calculated from the difference in the pendulum's period."

"If the skyscraper has an outside emergency staircase, it would be easy to walk up it and mark off the height in barometer lengths. If you wanted to be boring and orthodox, of course, you could use the barometer to measure the air pressure on the roof of the skyscraper and on the ground, and convert the difference into a height of air" (Abraham Pais 1991).

The student went on to say that since he is continually being urged to seek new ways of doing things, probably the best way would be to knock on the janitor's door and say: 'If you would like a nice new barometer, I will give you this one if you tell me the height of this building'." The student was allegedly Niels Bohr.

This might be an invented story, and Bohr may not have even been involved in this. The important point here is: there is no definitive way of thinking when answering a question. Creativity should be encouraged.

TIME IN MYTHOLOGY

At another level, we relate our illusive concept of time to death. In Sanskrit, Kala means time, and the Hindu mythology refers to Kala (Yama) as the personified God of death. It's no surprise that time and deaths are synonyms in the Puranas. The conversation between Nachiketa[2] and Yama is the theme of the *Kathopanishad*, one of the 108 Hindu scriptures of the Vedanta philosophy.

---

[2] Nachiketa is a character in the Upanishad and is noted for his rejection of material desires and his single-minded pursuit of the path of realizing Brahman.

Nachiketa says to Yama: "Some say that when man dies, he continues to exist, others that he does not. Explain, and that shall be my third gift."

Initially reluctant to explore such a subject matter, Yama engages in a deep and extensive conversation with Nachiketa and concludes: "The Self knows all, is not born, does not die, is not the effect of any cause, is eternal, self-existent, imperishable, ancient. How can the killing of the body kill Him?"

Did time begin with space at the Big Bang? Our current models of the universe emphasize this supposition, and it say should end with space. The second law of thermodynamics, often called the "death warrant of the universe," asserts that the entropy (disorder) of the universe must increase and thus the cosmological arrow of time must always go forward. Thus, cause precedes effect.

All physicists agree on the absence of universal time, because time is a matter of perspective. We can divide time periods into any unit from years to nanoseconds or even beyond. What is the smallest interval of time that can exist? Quantum theory says that it is Planck time, which is equal to 10^{-43} s. Below that, the meaning of time loses any physical meaning. So we have to say the universe came into existence when it was 10^{-43} s old. What if reality unfolded even at an earlier time? There are no answers, at least for now.

On grander scales, time should not be larger than 13.7 billion years, the current age of the universe. Yet, we do not know if time had the same rate of flow in every phase or part of the universe. This is why some physicists argue that the veil of illusion attributable to time must be removed to learn the truth. Theoretical computations show that the expansion of the universe at an accelerated rate may also be the result of our false notion of time. They anticipate a subsequent revolution in physics that would be a cosmos depicted without the constraint of time. But, what if the universe is itself an illusion followed by life?

Time is the most familiar experience we deal with, yet we don't know much about it. Black holes provide us with a clue to what one can call the "end of time." The intense gravitational force slows and ultimately freezes time. For an observer outside the black hole, a falling object would appear frozen in time.

"The black hole teaches us that space can be crumpled like a piece of paper into an infinitesimal dot, that time can be extinguished like a blown-out flame, and that the laws of physics that we regard as 'sacred,' as immutable, are anything but," wrote John Wheeler (1998), the visionary physicist who coined the term "black hole."

As one can imagine, the pursuit of the real nature of time or its illusion is profoundly complex. But that doesn't impede modern scientists from trying to crack the code. The latest attempt is by Sean Carroll, a Caltech physicist. His book, *From Eternity to Here: The Quest for the Ultimate Theory of Time*, published in January 2010, offers a lucid and fascinating discussion about the nature of time. He hypothesizes that our universe may be a relatively young member of a big family and that in several of our sibling universes time runs in the reverse direction. Some others, he argues, don't experience time at all; once a universe cools off and reaches maximum entropy, there is no past or present.

In the section "Aranya Parva" in the Mahabharata,[3] Yama disguised as Yaksha posed many questions to challenge Yudhisthira.

---

[3] The ancient religious epic of India.

Yaksha asks: "What is the greatest wonder in the world?"

Yudhisthira replies: "Every day, men see creatures depart to Yama's abode and yet, those who remain seek to live forever. This verily is the greatest wonder" (Rajagopalachari 1967).

So, what is time? We could say time is what a clock reads, but then we have clocks reading different times. Is it the greatest mystery? It is a mystery because we follow it yet we don't what it is exactly. However, the fact that we exist on a small rocky planet and are able to ask these questions are a greater mystery than time itself. We gift these questions to our future generations and hope they begin from where we stopped.

HOW TO DEFINE A YEAR?

Recently, a proposal by a joint task force of scientists to create a precise definition of *year* has sparked a debate among different segments of the scientific community. However, if one looks at these suggestions closely, it's obvious that the dispute stems from the inconsistent abbreviation of units that are in common use to represent a year. On the other hand, the attempts to build a universal definition of year or time go beyond the scope of these deliberations and have been unsuccessful since time immemorial.

Conventionally, we use year as the period of time required for Earth to make one revolution around the Sun, and it is written as y, yr, and in some scientific contexts as a, which is taken from the Latin word *annus,* meaning year. The higher magnitudes of year can be written as ky, kyr, or ka, for 1,000 years, and My, Myr, or Ma for 1 million years.

Whether one writes 10 Ma, 10My, or 10 Ma, they all convey the same interval between two events. But for the geologists, 10 Ma means 10 million years ago. If the new system is accepted, as proposed by the task force, it would eliminate the different conventions of writing these units; for example, in this case, it would be written as just 10 Ma instead of 10 My or 10 Myr. That will definitely bring the much-needed consistency among different branches of science when it deals with the concept of year or time.

Nonetheless geologists, who denote Ma for millions of years ago, complain they have no choice except to emphasize that point explicitly to imply the same meaning. But as a remedy, they can append a new notation for "ago," something like mya, which is already in use for "millions of years ago" in some disciplines. The other possible solution is to replace y, yr, or a with a new symbol for all. For astronomers and cosmologists, who often refer to future years such as 10 million years "after" or "from now," this whole discussion doesn't seem like a concern at all.

When we say 10 million years ago, it points to that amount of time "before present." The ambiguity in "present," as claimed by some, is insignificant, as the present is generally agreed to as 1950 CE, reflecting the fact that radiocarbon dating became practicable in the 1950s.

The need for standardizing different units of time is explicable and must be welcomed. The rather small inconvenience caused by it should not be a reason to ignore these suggestions for standardization. But, defining year or time is a completely different game – not an easy thing to do, as it appears.

The internationally accepted unit of time, second, is defined as the length of time required for 9,192,631,770 cycles of the Cesium atom at zero magnetic field. Thus the second defined was equivalent to a tiny fraction of the total amount of seconds in the chosen standard year 1900. As Earth's rotational time slows with respect to the atomic clock, leap seconds have to be inserted periodically to keep Earth's rotational time and the atomic clock time synchronized.

We use the Gregorian calendar, in which a year is exactly 365.2425 days, while astronomers use the Julian calendar that is adjusted to have 365.25 days in a year. They are in good enough agreement for a long time to come. However, it is easy to see how all these different time scales will cause problems. This is not unexpected and is a reflection of the intrinsic nature of time. The argument for a common definition of time lacks any meaningful insight.

On closer look, we can see that measuring and defining time are, in fact, not as different as we presume. Whether we use atomic clocks or natural clocks to measure time, we are not actually measuring time but defining time as "what a clock measures," as Einstein said. Rationalizing the units is of practical importance for several reasons and must be a priority, but defining time – don't even waste your time. In the grand scheme of things, some theoretical physicists argue, time may not even exist at all. We may have to begin the story of time as "once upon a time there was no time."

THE MYSTERY OF EXTRA DIMENSIONS

For us, space seems quite familiar as we experience it all the time and can move back and forth within it. However, on deeper analysis, the concept of space is equally or more challenging than the concept of time.

The startling assertion by world-renowned physicist Stephen Hawking and his co-author Leonard Mlodinow in their 2010 bestseller, *The Grand Design*, "Science can explain the universe without the need for a creator," generated worldwide headlines last year. Not as extensively reported, however, is the fact that for such a universe, or, even more remarkably, universes, to exist, we require 11 dimensions instead of the familiar four dimensions of space and time?

It is extremely difficult for us to comprehend the additional dimensions that we are precluded from observing. Our life is like a movie that unfolds on a 3D canvas. It is easy to demonstrate that we experience our world with three spatial dimensions. Einstein appended time as a fourth dimension, but even that is hard for non-physicists to envision. The three spatial dimensions are part of our daily life. We have the freedom to move back and forth, left and right, and up and down. Traditional mathematics describes such parameters as the X, Y, and Z coordinates. This is the degree of freedom we enjoy. Well, it turns out that we may be prisoners of this 3D reality, leading our lives in minuscule part in the higher-order universe, which exists in 11 dimensions.

We may visualize things in three dimensions, but that is no guarantee that our universe operates on those observable dimensions. A man walking on a tightrope has the freedom to move only back and forth and would describe the rope as a one-dimensional system, because, in mathematical terms, he enjoys the freedom to move along just one axis. On the other hand, a bug on the same rope has the freedom and available space to move sideways, besides back and forth. In its own perspective, for the bug, the rope is a two-dimensional system. Both are right, as their experiences are compatible with the observation of the structure on which they operate.

Our perception of reality is a based on 3D images that our brain creates, and to aid that process, it absorbs information from our surroundings, which is filtered through our sense organs. The models and laws we create are all based on that very same premise, that the external world operates in three dimensions. Is our brain capable of visualizing higher dimensional objects? Such queries demand a scientific search that connects physics with consciousness.

Fish in a pond or bowl, if equipped to describe their outside world with mathematical equations, would definitely have a different set of laws than ours. Strangely, their equations would be compatible with the observations they would make about the external world, and they would expect their theory of the universe to be right. They might even argue that this must be true for every observer. Is that true for humans as well? Just because we look at our universe and create a model that explains how it works rather than why it works doesn't mean we comprehend the complete picture. The fact is that our experience of the universe could be similar to that of the fish trapped in the bowl. We, fish, and possibly every other observer operate within the constraints of our physical dimensions.

In a 2005 article, Lisa Randall, a theoretical physicist at Harvard University wrote, "People have often made the mistake of believing only in what they could see. Extra dimensions might turn out to be one among many aspects of the cosmos about which we were initially mistaken... Given how extra dimensions – or whatever we discover – will tell us about the fundamental nature of our universe, do we have any choice but to explore?"

The idea that our four-dimensional space-time is embedded in some higher dimensional space is not really new in physics. Conversely, physicists' search for a theory of everything has landed them in the exotic worlds of 11 dimensions and many universes. What is equally shocking is that the so-called laws of nature, which we assume are the same everywhere in the universe, may not be so. The absolute laws of nature are giving way to laws that apply to particular universes. Additional dimensions and universes with their own sets of laws and constants are no longer a nightmare for physicists, who are growing increasingly comfortable with realities that defy our common sense observations.

Are the supreme truth and beauty of the cosmos revealed through mathematical equations by Einstein and Newton fading?

The absolute laws of nature are merely functionally effective, but imprecise, laws of convenience. There may be no absolute truth but only conclusions drawn from observations and premises.

A few years ago, there was a lot of enthusiasm among physicists who were enamored by string theory, in which the basic constituents were not tiny particles but strings. The one-dimension strings were believed to create cosmic vibrations in every possible way. The elegance of the string theory was admirable, but it faced profound theoretical and experimental challenges. The theory's popularity has diminished, and theoretical physicists are somewhat lost in the strange landscape of their own creation – at least so far due to the lack of any experimental or observational evidences.

Nonetheless, their pursuits have led them to a more exotic world in which one-dimensional vibrating strings are not the only foundation of the universe. A new picture of the cosmos is emerging that encompasses strings, particles, and even two-dimension membranes. This new paradigm, known as M-theory, is a network of all available possibilities. There is no agreement on what M stands for – membrane, magic, miracle. But it comes with a huge price; the theory demands eleven dimensions, ten of them spatial, for the universe and its forces to exist. In addition, it assumes a very large number of universes, both like and unlike ours, with their own laws.

However, if other spatial dimensions exist, why don't we experience them? One explanation is that these other dimensions are curled up and extremely tiny in our universe, making them invisible to human experience or our machines. "As of now, string theorists have no explanation of why there are three large dimensions as well as time, and the other dimensions are microscopic. Proposals about that have been all over the map," according to Edward Witten (Interview: Edward Witten, PBS 2003; http://www.pbs.org/wgbh/nova/elegant/view-witten.html), a pioneer in string theory and a professor at the Institute of Advanced Study, in Princeton, NJ.

Edwin A. Abbott's satirical novel, *Flatland: A Romance of Many Dimensions* (1992) used a fictional two-dimensional square to offer experiences of a whole new three-dimensional world with the help of a three-dimensional sphere. The science writer Isaac Asimov suggested that *Flatlands* was "The best introduction one can find into the manner of perceiving dimensions." Unfortunately, unless a four-dimensional creature lifts us from this 3D space land and grants us a magic view, like the square in *Flatland*, we will never know about other dimensions. The hope of finding their existence, if ever, rests with the ongoing high-energy experiments at the Large Hadron Collider in Geneva, Switzerland.

Researchers have also established experimentally that ghostlike particles, such as neutrinos, show up in our world from nowhere. Many believe that these particles come from a world beyond ours. The neutrino detector AMANDA (Antarctic Muon And Neutrino Detector Array), buried some 1.5 miles beneath the snow surface of the South Pole, is designed to capture neutrinos and is currently in action. Deeper study of neutrinos will provide support for the arguments that these particles, truly, are messengers from another world.

Some researchers are hopeful that at least some of the extra dimensions are big enough to measure, but that they might have shrunk as the universe expanded from its initial state. If so, some of the satellites deployed to map the picture of the early universe could reveal the faded signals of the extra dimensions.

The late physicist John Wheeler, who coined the term "black holes," once wrote, "In order to more fully understand this reality, we must take into account other dimensions of a broader reality."

The additional dimensions may hold the key to explaining the weakness of gravity, which is thought to leak into other dimensions, and such phenomena as dark matter and dark energy. The halo of dark matter that is assumed to be holding our universe together may partly exist in other dimensions than the ones we are used to.

The Grand Design explains that M-theory proposes a humongous 10^{500} apparent laws, each of which can potentially create their own universe, and ours happens to be just one of them. The model-dependent realism argues that from any given initial conditions and the ensuing chaos, a universe could emerge without the intervention of a divine being. The force of gravity can shape space, time, matter, and life, as dictated by the apparent laws of each universe. Just as a sheet of paper can be folded in many different ways, in fact, infinite ways, and each initial condition can give rise to a specific outcome, the force of gravity creates a multitude of universes depending on the existing initial conditions. This "something from nothing" is explainable by adding extra dimensions and many universes.

Our universe may be stuck in a three-dimensional membrane, making it impossible to detect other dimensions. Among all the known forces, theoretical physicists believe gravity could be an inter-dimensional force making its presence in all possible universes. We may be the inhabitants of this three-dimensional pocket called our universe, which is part of a higher dimensional universe.

However, if the extra dimensions exist, they would leave behind a signature in our universe that could be detected. One such signature would be the graviton, a particle that is supposed to be carrying the force of gravity. This yet-to-be-seen particle may be trickling into other dimensions, making gravity appear weaker here.

Many physicists agree that it would be nearly impossible to observe tiny extra dimensions. They associate this condition with atoms that we don't see in our daily life, but we know, without doubt, that they exist and make up our own world.

In his comedy science fiction, *The Restaurant at the End of the Universe*, Douglas Adams wrote, "There is a theory which states that if ever anybody discovers exactly what the universe is for and why it is here, it will instantly disappear and be replaced by something even more bizarre and inexplicable. There is another theory which states that this has already happened" (1980). Most do not concur with this satirical comment on extra dimensions, although it must be admitted that many physicists are increasingly perplexed by the concept of extra dimensions.

When it comes to nailing down the concepts of extra dimensions, the line between science and fiction becomes blurred. Conversely, it's purely logical to ponder our thoughts on such scenarios. Everyone agrees we can't hear all frequencies of sound nor can we see all wavelengths of electromagnetic spectrum, and it is a natural extension of that logic that we can't experience all dimensions.

On a deeper analysis, we are convinced and agree with what Albert Einstein concluded: "Reality is merely an illusion, albeit a persistent one." Definitely, extra dimensions are key players in the biggest puzzle nature presents to us.

REBIRTH OF TIME

We are repeatedly told that time is an illusion. Physicists consider time as an illusion mainly due to the fact that the laws of physics obey time-reversal symmetry. They seem to be symmetrical under time reversal, which means *there is no preferred direction for the flow of time*.

However, our daily experience is different than what the laws of physics proclaim. We do not remember the future but can remember the past. Our world and universe proceed in just one direction. Physicists explain this paradox on the basis of second law of thermodynamics, as briefly mentioned earlier. The entropy, which can be interpreted as the measure of disorder, of the universe is increasing, and the universe proceeds in the direction of that increase in disorder. There is no going back, at least in this universe.

In the meantime, the rebirth of time in physics could not be ruled out as some physicists have envisioned recently. Theoretical physicist Lee Smolin in his new book, *Time Reborn: From the Crisis in Physics to the Future of the Universe* (2013), emphasizes that we should take a fresh look at the whole concept of time again. Smolin affirms that "not only is time real, but nothing we know or experience gets closer to the heart of nature than the reality of time."

According to Smolin the long held notion, since the time of Plato, that time is not real has led us to a stalemate in physics. Smolin proposes an idea called *cosmological natural selection,* analogous to the evolutionary process in biology, in which the laws of nature evolve over time but not time. This may not be a completely new awareness, however, as in this scenario the time itself remains a constant. This hypothesis claims that, this time-bound evolutionary picture of the universe is best suited to the world as we observe it, because what we observe is in fact a universe evolving continually in time.

Both Newton and Einstein believed in the laws of physics that transcend the clutches of time and space. Their search for the ultimate reality or truth has somehow become the guideline for modern physicists, and it continues even now. They also had strong conviction and hope that 1 day we should be able to comprehend and express these laws of physics in the language of mathematics.

Now, a rethinking to this approach is gaining some momentum in modern physics, though slowly. As cosmological natural selection indicates there may not be any such laws of nature that can remain timeless. Like any other mechanism in our experience undergoes evolution, the laws of nature might also do the same over a large period of time. Truth and beauty, in absolute terms, are concepts of not just of physics but to a large extent have governed our thought process for a long time. These ideas were expressed in many forms such as the arts, literature and in philosophical discourse as well. So in our recorded human history we always pursued idealistic phenomena called absolute reality, assuming it exists without the boundaries of time and space.

Even if we acknowledge or accept the absence of such ideal value, this will not likely spread anarchy in science or society. But, it should be construed as a step closer to knowing our universe.

REFERENCES

Abbott, E. A. (1992). *Flatland: A romance of many dimensions.* New York: Dover.

Abiko, Seiya. (2000). Einstein's Kyoto address: 'how I created the theory of relativity'. Historical Studies in the Physical Sciences 31(pt.1): 1–35.

Abraham Pais. (1991). Niels Bohr's Times, In Physics, Philosophy, and Polity. First Edition. Oxford University Press, USA.

Adams, D. (1980). *The restaurant at the end of the universe: The hitch hiker's guide to the galaxy 2.* London: Pan.

"BBC – Religions – Hinduism: Vishnu." BBC – Homepage. http://www.bbc.co.uk/religion/religions/hinduism/deities/vishnu.shtml. Accessed 18 Aug 2010. 7. Whitehead. *Process and Reality, 163*(343), 375–77, 469.

Callender, C. (2002). Thermodynamic asymmetry in time. In: N. Z. Edward (ed.), *The Stanford encyclopedia of philosophy* (Spring 2002 edition). http://plato.stanford.edu/. Stanford: Stanford Encyclopedia of Philosophy.

Carroll, S. M. (2010). *From eternity to here: The quest for the ultimate theory of time.* New York: Dutton.

Craig, W. L. (2001). *Time and the metaphysics of relativity.* Dordrecht: Kluwer Academic.

Davies, P. C. W. (1995). *About time: Einstein's unfinished revolution.* London: Viking.

Eaves, E. (2011). "What is time? – Forbes.com." Information for the world's business leaders – Forbes.com. http://www.forbes.com/2008/02/28/what-is-time-oped-time08-cx_ee_0229thought.html. Accessed 4 Apr 2011.

Einstein, A. (1920). *Relativity: The special and the general theory* (pp. 23–27). New York: Nenry Holt.

Franck, F. (1976). *The book of Angelus Silesius* (1st ed.). London: Random House.

Gefter, A. (2008). *Is time an illusion?* http://www.newscientist.com/article/mg19726391.500-is-time-an-illusion.html. Accessed 15 Sept 2010.

Griffin, D., & Claremont Center for Process, S. (1986). *Physics and the ultimate significance of time: Bohm Prigogine, and process philosophy.* Albany: State University of New York Press, eBook Collection (EBSCOhost), EBSCO*host*, viewed 6 June 2013.

Hawking, S. W., & Mlodinow, L. (2010). *The grand design.* New York: Bantam Books.

Kiefer, C. (2007). *Quantum gravity* (2nd ed.). Oxford: Oxford University Press.

Mehra, J. (1994). *The beat of a different drum: The life and science of Richard Feynman.* Oxford: Clarendon Press.

Newsflash: Time May Not Exist | DiscoverMagazine.com. (2007). *Newsflash: Time may not exist | DiscoverMagazine.com.* Available athttp://discovermagazine.com/2007/jun/in-no-time#.UVCasFcuv3A. Accessed 25 July 2007.

NOVA | The Elegant Universe | Edward Witten | PBS. (2010). *NOVA | The elegant universe | Edward Witten | PBS.* Available at: http://www.pbs.org/wgbh/nova/elegant/view-witten.html. Accessed 25 Oct 2010.

Ono. (1982). *Physics today.* 45–47.

Pais, A. (1991). *Niels Bohr's times, in physics, philosophy, and polity.* Oxford: Clarendon Press.

Rajagopalachari, C. (1967). *Bhagavad-Gita* (3rd ed.). Bombay: Bharatiya Vidya Bhavan.

Rajagopalachari, C. (1968). *Mahabharata* (9th ed.). Bombay: Bharatiya Vidya Bhavan.

Randall, L. "Why i believe in higher dimensions – telegraph." *Telegraph.co.uk – telegraph online, daily telegraph, Sunday telegraph – telegraph.* http://www.telegraph.co.uk/technology/3341260/Why-I-believe-in-higher-dimensions.html. Accessed 23 June 2009.

Rovelli. C. (2008). *Forget time.* Available: http://fqxi.org/data/essay-contest-files/Rovelli_Time.pdf ?phpMyAdmin=0c371ccdae9b5ff3071bae814fb4f9e9. Accessed 12 July 2010.

Schweber, S. S. (1994). *QED and the men who made it: Dyson, Feynman, Schwinger, and Tomonaga.* Princeton: Princeton University Press.

Smolin, L. (2013). *Time reborn: From the crisis in physics to the future of the universe.* Boston: Houghton Mifflin Harcourt.

Sykes, C. (1994). *No ordinary genius: The illustrated Richard Feynman.* London: Weidenfeld and Nicolson.

The New York Times. *Past, present, and the quantity of the year.* May 4, 2011, New York edition.

Wheeler, J. A. (1968). Superspace and the nature of quantum geometrodynamics. In C. M. DeWitt & J. A. Wheeler (Eds.), *Battelle rencontres* (pp. 242–307). New York: Benjamin.

Wheeler, J. A., & Ford, K. (1998). *Geons, black holes, and quantum foam: A life in physics* (pp. 85–92). New York: W. W. Norton.

Whitehead, A. N. (1929). *Process and reality.* New York: Macmillan, 262.

Whitehead, A. N. (1968). *Modes of thought.* New York: The Free Press.

4

What's the Matter with Antimatter?

> *"Surely something is wanting in our conception of the universe. We know positive and negative electricity, north and south magnetism, and why not some extraterrestrial matter related to terrestrial matter, as the source is to the sink… Worlds may have formed of this stuff, with elements and compounds possessing identical properties without own, indistinguishable from them until they are brought into each other's vicinity… Astronomy, the oldest and most juvenile of the sciences, may still have some surprises in store. May antimatter be commended to its care! … Do dreams ever come true?"*
>
> –Sir Arthur Schuster, German-British physicist who coined the term for antimatter (1898)

The above quoted lines of Arthur Schuster reflect a purely imaginative prediction long before the 1932 discovery of the positron, the antiparticle of the electron (Anderson 1933). But, as you would see the antimatter is a reality now. So, where do we begin with?

The battle between good and evil is the primary theme of every mythology, religion, and folklore. Even modern human history is interpreted in a similar way, though it is often hard to recognize who is evil and who is virtuous. Probably, it is an inherent human engineering aspect that we wish to see everything in pairs with contrast. We have plenty of such pairs, as in good and evil, God and Satan – the list continues.

Not only have we graciously accepted the existence of these pairs in contrast, we also would like to see the good win over the evil. It seems like the most logical thought of humankind – science – is also not an exception. In particle physics, "Every particle has a corresponding antiparticle." While particles such as electrons, protons, and neutrons make

S. Mathew, *Essays on the Frontiers of Modern Astrophysics and Cosmology*, Springer Praxis Books, DOI 10.1007/978-3-319-01887-4_4, © Springer International Publishing Switzerland 2014

up the ordinary matter, their antiparticles, such as positrons, antiprotons, and antineutrons, make up antimatter. So it is logical to imagine the existence of any anti-entity such as anti-planet or anti-human. However, the existence of any such antimatter is ruled out, at least in this universe, because expectedly, yet surprisingly, the matter won over the antimatter in the battle of existence that occurred 13.7 billion years ago. The aftermath of that battle is what we call CMB (Cosmic Microwave Background) radiation, which we already have discussed in Chap. 1, establishing the Big Bang as the most acceptable theory for the origin of the universe.

The film *Angels and Demons,* based on Dan Brown's novel by the same name, was released in 2009. The story depicts the life of a secret group called the Illuminati. Portrayed in the background of the historic conflict between science and religion, the novel describes the efforts of the Illuminati to destroy Vatican City using antimatter, for revenging the atrocities carried out on them by the Catholic Church. Even though the existence of antimatter is proven, the biggest particle accelerators in the world have been able to create only a handful of antiparticles. In fact, CERN and Fermi Lab have produced antiparticles, including anti-hydrogen atoms, but not in any measurable amounts.

As of now, antimatter particles are considered to be the most expensive particles to produce, with the cost estimated to be $62.5 trillion a gram. Given the cost of production and the nonexistent storing techniques, these particles should remain almost as elusive, at least for the near future. Yet, the word antimatter provokes excitement in both science and fiction.

Generally, the laws of physics will only allow equal quantities of matter and antimatter to be produced in decay events (called "CP symmetry"–Charge conjugation Parity

Figure 4.1. Particles and antiparticles. Particles on the *left* from *top* to *bottom*: electron, proton, neutron. Antiparticles on the *right top* to *bottom*: positron, antiproton, antineutron (Source: Wikimedia Commons).

symmetry). However, there are some exceptions known within the standard model of physics. It has been known since 1957 that weak interactions violate both the C and P symmetries. The following section explains how we owe our life to this symmetry violation!

THE CHILDREN OF VIOLATION

In our everyday life, violations of enforced law result in tickets and can often advance to more complicated situations. However, physicists point to the violation of a physical law that took place billions of years ago that eventually led to the creation of everything, including us. We are essentially the children of that violation.

Physicists believe that at some point in the early moments of our universe a symmetry violation occurred, paving the way for matter to win over antimatter particles. As time progressed, this triumphant matter manifested in different forms such as stars, planets, life, and people. In that sense, we are obliged to this violation that prompted even our own existence.

The latest results from experiments provide clues to understanding this violation. Recently, scientists at the Relativistic Heavy Ion Collider (RHIC) reported the findings of their research carried out in the 2.4-mile particle collider located in Upton, NY. This "atom smasher" re-creates conditions similar to that which existed in the early universe. The data from the RHIC experiments fascinatingly points to a possible parity violation in strong interactions that binds the quarks and gluons.

Subatomic particles such as neutrons and protons are actually complex in nature. Both of these nucleons are composed of three quarks and gluons. At the RHIC experiment, quarks and gluons are ripped from their parent nuclei to create a hot soup of quark-gluon plasma. Physicists believe that matter existed in a physical state identical to this in the universe microseconds after the Big Bang.

Researchers re-create this condition at RHIC by using high-energy collisions of gold nuclei. The resulting temperature of their latest experiment measured 4 trillion degrees Celsius – the hottest temperature ever attained in the laboratory. The liberated quarks and gluons behaved more like a liquid, with considerable cooperation among them. This "new material" and their properties might be hiding the secrets of our cosmic origin.

As explained earlier, the standard model of particle physics demands a symmetry known as charge-parity (CP) to be conserved in the universe. The laws of physics must remain invariant for particles under charge exchange (C symmetry) or left and right swapping (P symmetry). This implies our world is indistinguishable from its mirror image.

The strong interaction that operates between quarks and gluons oppose parity violations under ordinary conditions. The data from the RHIC experiments intriguingly points to a possible parity violation in strong interactions. This cannot be perceived as breaking the known laws of physics, since these very same laws predicted such effects. Yet, it may be helpful to learn more about the earliest violation that shaped our cosmic history.

According to our current understanding, there existed equal numbers of particles and anti-particles in the early universe. Equal numbers of matter and antimatter particles completely annihilate each other to produce energy and conserve symmetry. Yet, we know that in our universe matter dominates antimatter.

If this symmetry had never been broken, the matter and antimatter would have completely annihilated each other, leaving the whole universe as pure energy. Though scientists will not be able to explain the global dominance of matter over antimatter with the information from their latest results, it will definitely assist them in probing that direction.

In fact, physicists celebrate when the laws of nature are violated. For when this happens, it provides physicists an opportunity to search for a deeper understanding of nature. As a result, new concepts and theories of physics can emerge.

The increased understanding attained about the quark-gluon interactions in the RHIC experiments will assist scientists in probing deeper and transferring the gained knowledge to LHC experiments at CERN, where much higher energy collisions are set to begin in coming months.

Once the imprints of broken symmetry become evident, we can reconstruct the puzzles of our early cosmic history. That would further underscore the significance of symmetry violation in the cosmic creation.

People often ask, "Why do we pump billions of dollars into these high-energy experiments? Who cares about quarks and gluons?"

The usual answer is "trying to answer the fundamental questions," which sounds like a cliché. That may seem justification enough to scientists but not necessarily to taxpayers.

The truth is that the technological spin-offs from these experiments are present in our everyday life, from superconducting materials used in maglev trains to PET scans in our hospitals – all share a common thread. Their origins can be traced back to these experiments.

Even the modern day World Wide Web was born in particle experiments. So let scientists "aim for the stars," and maybe we will reach the sky.

ANNIHILATION

The matter–antimatter annihilation produces pure energy in the most efficient way known to us. This remarkable energy formation from a nearly negligible amount of matter is the main cause of its appeal to the real world and also to fiction writers. According to Einstein's equation of mass-energy conversion, the mutual annihilation of 1 kg each of matter and antimatter can produce the energy equivalent to that of the most powerful nuclear bomb ever tested, that in turn uses hundreds of kilograms of nuclear material. This claim is based on the experimental evidence that the electron and its antiparticle positron annihilate each other, resulting in the production of energy, mainly in the form of gamma rays.

However, the mutual annihilation of heavier particles such as protons and its antiparticle (antiproton) is more complicated, as it produces exotic particles such as neutrinos along with energy production. Contrary to popular belief, the energy production of matter–antimatter annihilation is a more complex phenomenon than the relatively straightforward annihilation of an electron-positron pair.

The practical complications involved in harnessing the energy from the matter–antimatter annihilation have not dispirited scientists in their hunt for this rather magical energy. In fact, NASA has been actively pursuing this technique of energy production as a possible future rocket fuel. They envision an unlimited supply of energy from such a small

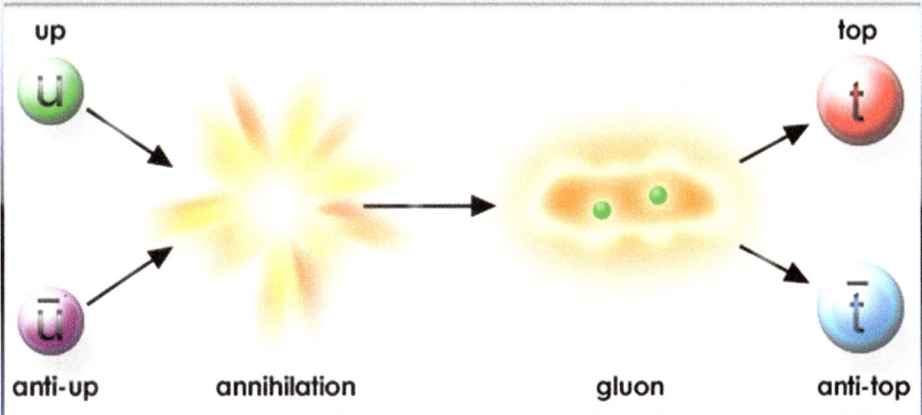

Figure 4.2. Particles and antiparticles annihilate each other. The diagram shows an up quark and an anti-up quark annihilating each other to create energy. From that energy another particle pair emerges. Creation and destruction are two sides of a coin (Image credit: Particle Data Group of Lawrence Berkeley National Laboratory).

amount of material that could propel the rockets to planets and stars, much beyond our destinations of the past. The future Mars mission plan of NASA would depend on this, though currently it is a hypothetical situation.

Though everyone agrees on the existence of the antiparticles now, this was not the case 80 years back, when it was predicted for the first time. Paul Dirac, one of the pioneers of quantum physics, who had conducted extensive study on the mathematical equations associated with the existence of fundamental particles, came upon a rather strange situation. Dirac predicted (1928) the possibility of an anti-world similar to that of our world but made up of antiparticles, a new radical idea in particle physics. Two years later, Carl Anderson (1933) at the California Institute of Technology, while examining the tracks created by cosmic rays, confirmed the existence of positrons – the antiparticle of electrons. However, it took 22 years of investigation to discover the antiproton, and 4 years later, antineutrons were found.

It was in 1995 that the first anti-hydrogen atom was created at the CERN, executing the knowledge already gained through earlier experiments conducted at the Fermi lab. It is assumed that the coming years will see the production of antimatter in larger quantities along with the complicated storage technique of this mysterious matter. This century, hopefully, might witness a spaceship being propelled to distant planets using the energy created by the matter–antimatter annihilation – something similar to that of the starship *Enterprise.*

The standard cosmology theory suggests that the universe began with equal amounts of matter and antimatter, like the twins of the Big Bang. However, in the early universe, this super symmetry was broken, for unknown reasons, which led to the dominance of matter in this universe. This is one of the unsolved mysteries in science, and is known as the baryon asymmetry problem in physics. Assuming that there were equal amounts of matter and antimatter in the early universe, their mutual destruction would have caused the complete absence of any matter in the universe. The whole universe would have been filled with energy (light) created in that process. The fact is that matter dominates in our universe, and there is hardly any evidence of antimatter present except when it is created by cosmic rays or in particle accelerators.

One possible solution to this puzzle is that there exists a universe with matter dominance (our universe), and another with antimatter dominance, without any interaction between them. This explanation is unlikely, however, as the scientific research of the last decades offers no solutions in that direction. None of the existing theories are capable of providing an explanation of the matter supremacy in our universe. The only optimism at this time is the Large Hadron Collider at CERN, which will shed some light into the broken symmetry that happened in the early universe, when it will operate again in coming months.

As the infant universe began its existence 13.7 billion years ago, the four fundamental forces separated from the super forces that existed prior to that event. The hot and dense universe inflated and cooled for a fraction of a second. This was followed by the radiation era, when particles and antiparticles annihilated each other to create radiation (photons). These photons, the "relic" of the Big Bang, coupled with matter, were unable to escape since they were trapped in the universe. The battle of existence continued between matter and antimatter, and that created more photons.

At some point, the matter particles exceeded the antimatter particles about one part per billion. As the dominant matter particles became stable, they lost interest in the ongoing coupling with the photons, which were born out of the battle waged between matter and antimatter. Now, these photons could travel through the expanding universe bearing the mark of the violent history of the early universe. These photons, currently existing in the microwave region of the electromagnetic spectrum known as CMB, were finally detected by Penzias and Wilson in 1964 as discussed in Chapter 1.

This was the battle won by the matter that caused the creation of stars and planets – the cause of our own creation. As in recorded human history the battle waged between two rivals gave rise to the creation of a new realm, the first ever battle between matter and antimatter gave way to the universe as we know it, so we call it "Mother of All Battles."

THE MOTHER OF ALL BATTLES

Modern science has shown that matter and the material world is just one side of a coin. If our world is made of normal matter, it is equally logical to think of a world composed of antimatter.

Matter and energy are manifestations of the same underlying entity. In particle physics, every particle has a corresponding antiparticle. Particles – like electrons, protons, and neutrons – make up ordinary matter, while their antiparticles – such as positrons (anti-electrons), antiprotons, and antineutrons – make up antimatter. Sequential logic allows us to imagine the existence of anti-entities, such as anti-planets or anti-humans. However, the existence of antimatter is ruled out, at least in this universe, because matter won over antimatter in the battle of existence that occurred 13.7 billion years ago in the Big Bang. The aftereffect of that battle is the Cosmic Microwave Background (CBM) radiation, which is the first energy (light), formed in the universe as a result of the mutual annihilation of matter and antimatter.

Similarly, the process of creation and destruction go continuously in the universe. The Vedic sages envisaged this process as a cosmic dance of creation and destruction through symbols like Nataraja, the dancing God Shiva. Death of an object is in fact a state change and the root cause for a new birth. As explained in the previous chapter this is most evident in nebulae, where new stars are born from the huge columns of dust and gas.

Now, it is an established fact that the antiparticles exist. But what good are they anyhow?

People often wonder about the futility of research that seems to transcend the landscape of our everyday life. The technique called Positron Emission Tomography (PET) is an example of the electron-positron annihilations. This practically beneficial technique is being used to reveal the workings of the brain or other organs in medical diagnosis. However, this annihilation has to be performed at lower energies to be appropriate for medical uses.

The matter–antimatter annihilation serves as another example of mass-energy conversion. As we understand, when a positron and an electron annihilate, all their mass is converted to energy, and given off in the form of gamma rays.

The positron required for the destruction of an electron comes from the radioactive nuclei incorporated in a special fluid injected into the patient. The positrons then annihilate with electrons in nearby atoms. The gamma rays emerging from this process are detected and analyzed to create the images of the inner organs.

If the antiparticles are real, one could imagine that there are objects of larger magnitude that are made of antiparticles, such as anti-people or anti-worlds similar to our familiar world that is the manifestation of particles. Our world will end in flash of light if it encounters an anti-world. As of now we don't know anything like that exists, at least in our universe.

Our classical depiction of the dual nature of existence is nothing new. This physical possibility of a dual world has expressions in mythology, religion, and folklore as the epic battle between good and evil. In Hindu literature, the Devas[1] govern the regions of heaven and are opposed by the demonic Asuras.[2] In the cosmic struggle between the forces of order and chaos, Devas support humans, who inhabit Earth. The conflict between Devas and Asuras is described in the myth of the churning of the "ocean of milk" (Palazhi) for treasures, including Amritha,[3] which provides strength (energy) and immortality.

Similar to the fight between good and evil portrayed in mythology, the battle between matter and antimatter is an ongoing event, although science has not yet confirmed the existence of such conflicts beyond the early universe.

Standard cosmological theory suggests that the universe began with equal amounts of matter and antimatter, like the twins of Big Bang. However, in the early universe, this super symmetry was broken for unknown reasons, leading to the dominance of matter in

---

[1] Good spirited benevolent supernatural beings inhabiting the heavens.

[2] Evil spirited materialistic beings, and they occupy the lower planes of the world.

[3] Referred to as a kind of nectar that provides immortality.

this universe. It is one of the unsolved mysteries in science, known as the baryon asymmetry problem in physics. Assuming that there were equal amounts of matter and antimatter in the early universe, their mutual destruction would have caused the complete absence of any matter. The whole universe would have been filled with energy (light) created in that process.

Again going back to the ancient Hindu text, the Rig Veda, one can find expressions on the dual nature of the universe as it describes the creation of the universe from the remains of a gigantic primeval Cosmic Man. Creation happened gradually; the universe in its primitive form primarily spread homogeneously throughout the universe. The complete equilibrium and homogeneity, when broken, led to an inhomogeneous state of the primordial fluid.

The Veda refers to primeval couples such as Heaven and Earth and celebrates their togetherness:

> Which of these two came earlier, which came later?
> How did they come to birth? Who, O Seers, can discern it?
> They contain within them all that has a name,
> while days and nights revolve as on a wheel.
>
> – (Shah 2006)

We, in our own scientific quest to make sense of this universe, ponder the same questions in a different way. Yet, both science and mythology fundamentally resemble the same human ambition to comprehend the unknown.

Before we think of the energy efficiency of matter–antimatter annihilation, let's think about how the already known and used methods of matter-energy conversion take place.

The power of matter-energy transition was vividly on display in the detonation of the nuclear bomb in the New Mexico desert in 1945. On seeing the fireball and mushroom cloud, J. Robert Oppenheimer, the director of the Manhattan Project, who had a deep interest in religion, Hinduism in particular, recalled a passage from the Bhagavad-Gita: "I am become death the destroyer of worlds. If the radiance of a thousand suns were to burst at once into the sky, that would be like the splendor of the mighty one."

The matter–antimatter annihilation produces pure energy even more efficiently than nuclear fission or fusion. However, it would be useful to digress here to learn more about nuclear fusion before we proceed with the matter–antimatter discussion.

WAITING FOR A STAR BIRTH

It took only a decade from the first atomic bomb to build nuclear fission reactors, which currently provide 15 % of the world's electrical energy. It's more than half a century since the first hydrogen bomb was experimentally detonated, yet we are far away from a fusion reactor.

If fusion is the most efficient mode of energy creation, stars are the best fusion reactors as we know them. So, making a self-sustainable fusion reaction on Earth is akin to creating a star on Earth. Scientists around the globe have been trying the same for decades with little success, often ridiculed by the public: "Fusion energy is always few decades away."

Are we on the verge of making this source of energy a reality? Given the history of fusion research, this might seem simply an optimistic consideration. However, if the strong signals from National Ignition Facility (NIF) are any indication, there could be an actual "star birth" that the scientific world has been waiting for.

The National Ignition Facility (NIF) at Lawrence Livermore National Laboratory in California was dedicated in May 2009. It houses the largest and strongest energy laser on the planet, and the facility began firing laser beams onto targets in June 2009. The goal is to fuse isotopes of hydrogen to generate carbon-free energy that will feed the electric grids.

There are many challenges to initiate and sustain a thermonuclear reaction. First of all, a typical deuterium-tritium[4] fusion needs a temperature above 100 million degrees Kelvin in the absence of gravitational compression, the kinds of temperatures at which the matter transforms into its fourth state, plasma, as it exists in stars. Deuterium, also known as heavy hydrogen, is an isotope of hydrogen, with one proton and one neutron, while tritium, another isotope, has one proton and two neutrons in its nuclei. Their fusion yields the next higher element in the periodic table – helium – along with the release of a tremendous amount of energy.

Secondly, the hot plasma needs to be confined long enough to sustain the fusion reaction. Generally, two methods are employed to achieve this. One is magnetic confinement, originally designed by the Soviets in the 1950s. A doughnut-shaped device called a tokomak is used to hold the plasma with its intense magnetic field. In fact, most of the researchers around the world still rely on this technique, including the development of the International Thermonuclear Experimental Reactor (ITER), which was launched in 2006 as an international collaboration. The construction of ITER, the world's largest science experiment, has been continuing in Cadarache, France.

NIF adopts a different technique in plasma confinement, known as inertial confinement, and hopes to hit the jackpot before ITER does. A powerful beam of 192 lasers is used to heat and compress the fuel pellet of deuterium and tritium quickly so that the imploding pellet keeps the material confined. Thus the inertia of the material is exploited in plasma confinement instead of using an external force. Last November, NIF announced that laser beams can be effectively delivered and are capable of creating sufficient energy to drive fuel implosion, an important step toward the ultimate goal of fusion ignition.

On January 27, 2010, the National Nuclear Security Administration (NNSA), the umbrella organization under which NIF operates, announced that scientists have successfully focused a record level of laser energy – more than 1 MJ – to a target in a few billionths of a second. This is about thirty times more than the energy used by any such experiments in the world, and this exhibits the capability to create conditions necessary for igniting fusion.

"Breaking the megajoule barrier brings us one step closer to fusion ignition at the National Ignition Facility and shows the universe of opportunities made possible by one of the largest scientific and engineering challenges of our time," commented NNSA administrator Thomas D'Agostino (Press release, http://www.nnsa.energy.gov/mediaroom/pressreleases/01.27.10 2010).

---

[4] Both are isotopes of hydrogen.

Figure 4.3. Stars are born from gas and dust. They burn hydrogen to heavier elements through fusion. If we can achieve controlled fusion on Earth, it will be the best solution for our energy crisis. Humanity's wait for a star birth on Earth has not been accomplished yet (Image credit: ESO).

This has certainly brought the NIF close to achieving nuclear fusion. In the past, focusing the laser energy effectively on tiny targets was a big challenge. With the mega-joule shot, scientists are gearing up for another stage. The next step is to ignite fuel capsules that require the fuel to be in a frozen hydrogen layer (at 425°F below zero) before the attempt on the actual fuel. The successful fusion at NIF will be a precursor to LIFE – the Laser Inertial Fusion Engine – a hybrid technology of fusion and fission being developed at Lawrence Livermore Lab, which may be years away.

In this era of global warming, fusion plants have another big advantage to offer. The fusion reaction does not produce any greenhouse gases. Also, the required fuels for fusion are readily available – deuterium can be obtained from water, and tritium can be bred in the reactor once it is self-sustainable. Finally, the birth of the star on the planet will help us to deviate from the habit of depending on the planet for energy needs, truly a remarkable transition for a civilization.

TRAPPING THE ANTIMATTER

Scientists have been studying antimatter particles in detail to provide us with clues to the long held secrets of the early universe. In 2011 (Andresen et al.), it was reported that antimatter particles (anti-hydrogen atoms) had been trapped for a record time of about 16 min.

Physicists have long wondered about the gravitational interactions between matter and antimatter. Even though the existence of antimatter is well established, the biggest particle accelerators in the world have been able to create only a handful of antiparticles. As mentioned earlier, CERN and the U. S. government's Fermi National Accelerator Laboratory (Fermilab), located in Batavia near Chicago, Illinois, have produced antiparticles, including anti-hydrogen atoms, but not in any measurable quantities But scientists are hopeful that in the coming years they will find ways to generate antimatter in larger quantities and develop advanced techniques to store this mysterious matter.

Because hydrogen atoms are the simplest in structure (one proton and one electron), its antimatter, anti-hydrogen (one antiproton and a positron), is the easiest antimatter to synthesize. The experiment captures antiprotons and combines them with anti-electrons (positrons) to make anti-hydrogen atoms. These are stored and studied for a few seconds in a magnetic trap. The ATHENA (AnTiHydrogEN Apparatus) at CERN was an international collaboration to produce anti-hydrogen atoms at low temperature for experimental purposes. ATHENA has produced tens of thousands of anti-hydrogen atoms.

However, it was practically impossible to capture the anti-hydrogen atoms because of the high temperatures associated with them. The high-energy anti-hydrogen atoms hit the wall of the experimental apparatus and annihilated with atoms of normal matter. The analysis of anti-hydrogen atoms would reveal the difference between matter and antimatter beyond theoretical assumptions and possibly validate the theories of the absence of antimatter in our universe.

Though the ATHENA experiment came to an end in 2004, another ATRAP (Anti-hydrogen Trap) collaboration is still in operation. Its goal is to create anti-hydrogen at a lower temperature that can be trapped long enough to compare with ordinary hydrogen.

The ALPHA collaboration, the successor of ATHENA, has recently (2013) reported a novel methodology to test the gravitational effects of antimatter. In the past, there were only indirect indications that there is no difference between gravitational attractions, whether it is matter–antimatter or matter-matter interactions. The latest experiment does not confirm any earlier indirect test results, but is considered as the first step towards answering more fundamental questions, such as antimatter and antigravity.

Does antimatter falls down, like normal matter, or does it fall up? Such questions should be answered in coming years as more precise experiments are used to nail down the mystery shrouding these strange particles.

Further projects are underway at CERN to conduct direct experimental tests of gravity on anti-hydrogen, and are now in their final construction phase. A propensity for anti-hydrogen atoms to fall downward when released from the ALPHA anti-hydrogen trap lead toward a definitive answer to the fundamental question of whether matter falls up or down.

The concept of antiparticles was inspired by mathematical reasons rather than physical. The Dirac equation, which conceived the idea of the positron, called for two solutions, one positive and one negative. For many, it was hard to believe that negative energy would someday be represented by a real particle with physical form. Yet, Dirac was determined to accept the implications of the solution which, in fact, represented both halves of reality.

As an analogy, imagine that negative numbers are not available in arithmetic operations. In such a scenario, what is 1–2 equal to? Fortunately, we do consider negative numbers as real numbers. So +1 and −1 are attuned with a difference. They are located on either side of zero at equal distance from the zero. Here, we have the benefit of dealing with a mathematical solution and do not need to carry the burden of assigning a physical meaning. But for Dirac, if his solutions are right, there must exist a positive electron. Thus a pure mathematical solution envisioned by Dirac, was later on realized by the discovery the positron. Now, we know that for every particle there is an opposite (charge) particle with the same mass, similar to positive and negative numbers in the real number system.

As we understood it, positrons are the result of theoretical considerations and a mathematical necessity. Ironically, there are practical applications that can come out of these elusive particles. For instance, the brain scanning procedure known as PET (Positron Emission Tomography) exploits the electron-positron annihilation to reveal the workings of the brain. The positron created by the radioactive decay process is used to annihilate an electron in an atom of the brain, rendering an image of the brain on the screen. Researchers are optimistic about the use of antiprotons in tumor irradiation in the future.

In the near future the researchers are hopeful that with more detailed experiments that trap anti-atoms for a longer time, they can explore the strangeness of antimatter to its full extent. If and when they succeed, it will be a tribute to the invocation by Paul Dirac, a pioneering quantum physicist, who predicted the possibility of antiparticles for the first time in 1928. "Pick a flower on Earth, and you move the farthest star."

SEARCHING FOR ANTIMATTER IN THE UNIVERSE

The hunt for antimatter has taken a celestial turn recently. The following paragraphs describe that exploration briefly.

The space shuttle *Endeavour's* final mission was remarkable in many ways. It was commanded by astronaut Mark Kelly, the husband of congresswoman Gabrielle Giffords, one of the 2011 Arizona shooting victims. This mission carried one of the most expensive payloads to the International Space Station (ISS) – a $2 billion detector called the Alpha Magnetic Spectrometer (AMS-02). The main goal of this experiment was to probe the unknown universe by searching for antimatter particles.

Physicists believe an asymmetry that occurred in the early universe is the root cause of everything, including our own existence. Theories suggest that equal amounts of matter and antimatter should have been created at the Big Bang. When they interact, both matter and antimatter completely annihilate each other in a flash of radiation. However, in the early universe, matter somehow dominated the antimatter, giving way to a universe filled with stars, planets, and galaxies.

The victory of matter over antimatter is a mystery. Where has the antimatter gone? Does a mirror image of our universe exist that's made of antimatter? Could there be isolated systems of antimatter in the universe? We don't have definitive answers to these questions at this point.

Though the laboratory experiments created antimatter particles and atoms in small quantities, cosmologists remain speculative about their large presence in the universe as Paul Dirac imagined. AMS, now installed on the long arm of the ISS, will collect the data to find these lost seeds of the Big Bang, if there are any left!

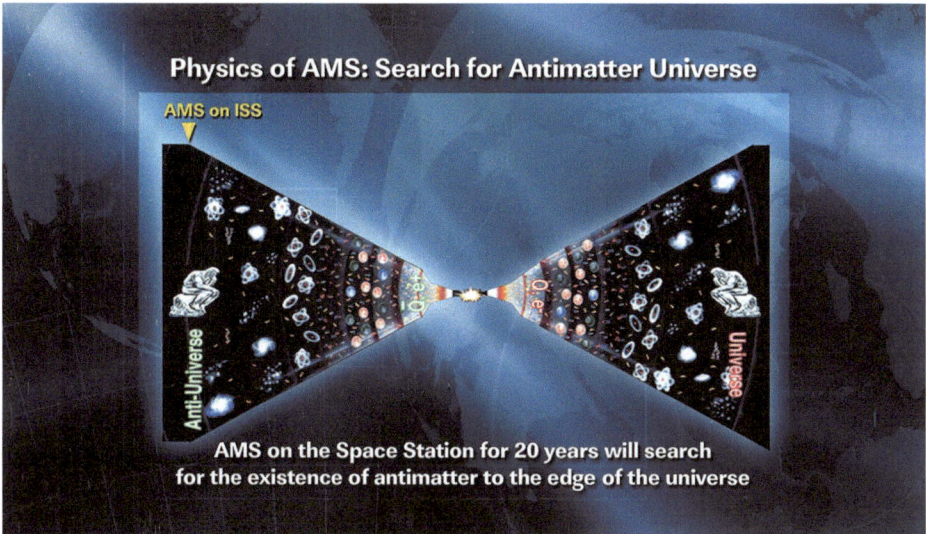

Figure 4.4. Antimatter particles, like matter particles, could exist in the form of a universe. Where is the antimatter universe? (Image credit: NASA).

The AMS observation will, in some way, support the idea that dark matter could consist of exotic particles that annihilate each other and create detectable cosmic rays made up of electrons and their antimatter particles, positrons. If the number of positrons at high energies suddenly plummets, that could be a signature of the elusive dark matter in the universe.

With all the features and possibilities associated with anti-particles, the scientific world is not convinced about the presence of a significant amount of antimatter in the universe. Finding the antimatter will help us to answer one of the fundamental questions on the origin and nature of our universe. Our quest for the deeper understanding of the intricate history of the universe and its complex composition will be revealed, if we can detect antimatter in considerable amounts in the universe. This speaks to the relevance of AMS and the search it has been conducting.

Antimatter particles, born out of equations and imagination, once again illustrates that human imagination is the most vital element of success in pursuing reality.

MORE ON PAUL DIRAC

Let us digress a bit here to sketch the life of Paul Dirac, who attempted to bridge quantum theory and the special theory of relativity and, in the process, provide a reason for suspecting that there is such a thing as anti-matter. In the book *The Strangest Man: The Hidden Life of Paul Dirac, Mystic of the Atom,* Graham Farmelo (2011) reveals the legendary personality who mostly spoke with one-word answers.

Paul Dirac was born on August 8, 1902, in Bristol, England. He acquired a degree in electrical engineering at the University of Bristol in 1921. He later on went to St. John's College, Cambridge, and received a doctorate in 1926, at which point he became the Lucasian Professor of Mathematics at Cambridge. "If Newton was the Shakespeare of British physics, Dirac was its Milton," described *The Telegraph* in 2009 in a review of Farmelo's book on Dirac.

When Albert Einstein was asked who he wanted to join him at the Institute for Advanced Study in Princeton, New Jersey, he named Paul Dirac. Paul Dirac won a Nobel Prize in Physics in 1933, at the age of 31, "for the discovery of new productive forms of atomic theory" along with Ervin Schrödinger. Dirac's work could be considered a productive reconciliation between the two theories, relativity and quantum mechanics.

Dirac was known among his colleagues for his precise and clever insights. Once Paul Dirac came to Oppenheimer and said: "Oppenheimer, they tell me you are writing poetry. I do not see how a man can work on the frontiers of physics and write poetry at the same time. They are in opposition. In science you want to say something that nobody knew before, in words which everyone can understand. In poetry you are bound to say… something that everybody knows already in words that nobody can understand." (Source: http://www.dirac.ch/PaulDirac.html)

Paul Dirac strongly believed that it was more important to have beauty in one's equations than to have them fit experimentally. The strangest man who predicted the strangest particles died on October 20, 1984, in Tallahassee, Florida, at the age of 82.

Dirac was the youngest theoretical physicist to receive the Nobel Prize in 1933. Though it was unpleasant for many physicists to depart from the determinism of classical physical theory advocated by Newton, it was becoming the new reality in those days. Paul

Figure 4.5. Paul Dirac (Image source: Wikimedia Commons).

Dirac accelerated the process of incorporating mathematical beauty in physics, and with the prediction of the existence of anti-particles, he accelerated the process of challenging the conventionalism that dominated physics. He never worried too much about the collapse of traditional views of physics. Physics in the future could be different than what we know today; it has always been like that. The so-called constants of nature and laws of nature may evolve over time.

The following paragraph narrates briefly the emancipation of Dirac's thoughts as appeared in *Scientific American* magazine in 1963. Even after half a century it is still a relevant piece to revisit.

It seems to be one of the fundamental features of nature that fundamental physical laws are described in terms of a mathematical theory of great beauty and power, needing quite a high standard of mathematics for one to understand it. You may wonder: Why is nature constructed along these lines? One can only answer that our present knowledge seems to show that nature is so constructed. We simply have to accept it. One could perhaps describe the situation by saying that God is a mathematician of a very high order, and He used very advanced mathematics in constructing the universe. Our feeble attempts at mathematics enable us to understand a bit of the universe, and as we proceed to develop higher and higher mathematics we can hope to understand the universe better.

It is fair to admit that we have reached a certain stage in physics at the present time, but it is not the final stage. It is just one stage in the evolution of our picture of nature, and we should expect this process of evolution to continue in the future, as biological evolution continues into the future. The present stage of physical theory is merely a stepping stone toward the better stages we shall have in the future. Dirac was optimistic about the future of physics and in our understanding of the universe. This view is reflected in his own words, "One can be quite sure that there will be better stages simply because of the difficulties that occur in the physics of today."

REFERENCES

Anderson, C. D. (1933). The positive electron. *Physical Review, 43*(6), 491–494.

Andresen, G. B., et al. (2008). Production of antihydrogen at reduced magnetic field for anti-atom trapping. *Journal of Physics B: Atomic, Molecular and Optical Physics, 41*, 011001.

Andresen, G. B., et al. (2011a). Confinement of antihydrogen for 1,000 seconds. *Nature Physics, 7*, 558–564. Available at: http://www.nature.com/nphys/journal/v7/n7/full/nphys2025.html. Accessed 08 Nov 2011.

Andresen, G. B., et al. (2011b). Search for trapped antihydrogen. *Physics Letters B, 695*, 95–104.

Andresen, G. B., et al. (2012). Antihydrogen annihilation reconstruction with the ALPHA silicon detector. *Nuclear Instruments and Methods in Physics Research, 684*, 73–81.

Dirac, P. A. M. (1928). The quantum theory of the electron. *Proceedings of the Royal Society A, 117*, 610–624. doi:10.1098/rspa.1928.0023.

Dirac. P. A. M. Nobel lecture (December 1933).

Dirac biography. (2012). *Dirac biography*. Available at: http://www-groups.dcs.st-and.ac.uk/history/Biographies/Dirac.html. Accessed 18 June 2012.

Dirac, P. (2013). *Paul Dirac*. Available at: http://www.dirac.ch/PaulDirac.html. Accessed 18 June 2013.

Farmelo, G. (2011). *The strangest man: The hidden life of Paul Dirac, mystic of the atom*. New York: Basic Books. First Trade Paper Edition.

First atoms of antimatter produced at CERN|CERN press office. (2012). First atoms of antimatter produced at CERN | CERN press office. Available at: http://press.web.cern.ch/press-releases/1996/01/first-atoms-antimatter-produced-cern#. Accessed 18 Feb 2012.

Flesher, P. V. M. (2011). Hinduism glossary. Default title. http://uwacadweb.uwyo.edu/religionet/er/hinduism/hglossry.htm. Accessed 18 Mar 2011.

Grant, A. (2013). Atom & cosmos: Cosmic rays hint at dark matter: Space station-based instrument confirms past experiments. *Science News, 183*(9), 14. Academic Search Premier, EBSCOhost, viewed 2 March 2012.

Hawking, S. W. (1988). *A brief history of time: From the big bang to black holes*. Toronto: Bantam Books.

Krantz, S. (2005). *Mathematical apocrypha redux: More stories and anecdotes of mathematicians and the mathematical (spectrum)*. Washington, DC: The Mathematical Association of America.

National Nuclear security Administrator (2010). NNSA announces unprecedented 1 Megajoule laser shot at the National Ignition Facility. Jan 27 2010.

Schuster, A. (1898). Potential matter-a holiday dream. *Nature, 58*(1503), 367.

Shah, R. C. (2006). *Ancestral voices – Reflections on Vedic, classical and bhakti poetry* (p. 14). New Delhi: Motilal Banarsidass.

Smith, H. (1991). *The world's religions: Our great wisdom traditions*. San Francisco: Harper San Francisco.

Smolin, L. (1997). *The life of the cosmos*. Oxford: Oxford University Press.

Smolin, L. (2006). *The trouble with physics*. Boston: Houghton-Mifflin.

The ALPHA Collaboration, & Charman, A. E. (2013). Description and first application of a new technique to measure the gravitational mass of antihydrogen. *Nature Communications, 4*, 1785. doi:10.1038/ncomms2787.

The Evolution of the Physicist's Picture of Nature|Guest Blog, Scientific American Blog Network. (2013). The evolution of the physicist's picture of nature|guest blog, scientific American blog network. Available at: http://blogs.scientificamerican.com/guest-blog/2010/06/25/the-evolution-of-the-physicists-picture-of-nature/. Accessed 18 June 2013.

Tsan, U. (2012). Negative numbers and antimatter particles. *International Journal of Modern Physics E: Nuclear Physics, 21*(1), 1250005-1–1250005-23. Academic Search Premier, EBSCOhost, viewed 1 June 2012.

5

Playing with Light

> *"Nature and Nature's laws lay hid in night;*
> *"God said Let Newton be! and all was light."*
>
> —Alexander Pope (1688–1744), British satirical poet,
> in the epitaph intended for Sir Isaac Newton's tomb
> in Westminster Abbey (1730).

WHAT IS LIGHT?

In the last chapter we have seen in detail the birth of antimatter through the work of physicist Paul Dirac. Dirac had an amazing ability to produce fundamental equations – the poems of science.

It has been said about him that when he was asked what an electron is, he wrote an equation on the board and said, "Electron, hmmm, Here, this is electron (What's light, 1999)."

This equation about light is well known and is used extensively in dealing with light:

$$E = h\nu$$

The Greek letter ν (nu) stands for the frequency of the light wave; c is the symbol for the speed of light; and h stands for a fundamental constant of nature known as Planck's constant.

In 1900s, the German physicist Max Planck worked extensively on the relation between the radiation an object emits is and its temperature. The result of his work could be stated as the above formula. This formula agreed very closely with the experimental data available in those days. But, it demanded a huge shift about how scientists perceived the nature of energy.

The Planck formula assumed that the energy of a vibrating molecule was quantized. This means energy cannot have all the values as thought previously but only certain values. Again, as the formula shows, the energy is proportional to the frequency of vibration.

S. Mathew, *Essays on the Frontiers of Modern Astrophysics and Cosmology*, Springer Praxis Books, DOI 10.1007/978-3-319-01887-4_5, © Springer International Publishing Switzerland 2014

Figure 5.1. Statue of Max Planck outside Humboldt university in Berlin, Germany. Max Planck made a huge effort to solve the mysteries of nature. But in his own words, "science cannot solve the ultimate mystery of nature. And that is because, in the last analysis, we ourselves are part of nature and therefore part of the mystery that we are trying to solve" (Source: Wikimedia Commons).

In other words energy comes in discrete packets called "quanta" or "chunks," and it can be calculated by multiplying frequency with a certain constant. This constant came to be known as Planck's constant, or h, and it has the value 6.626×10^{-34} J s.

This work has actually laid the foundation for quantum mechanics and was considered as Planck's most important contribution to the new physics that emerged in the early twentieth century. It is also considered as a turning point in the history of physics. Initially, there was immense resistance to Planck's work. However, later on, the evidence for its legitimacy became overpowering in many areas of physics. This new approach accounted for many discrepancies between observed phenomena and classical theory.

THE ELECTROMAGNETIC SPECTRUM

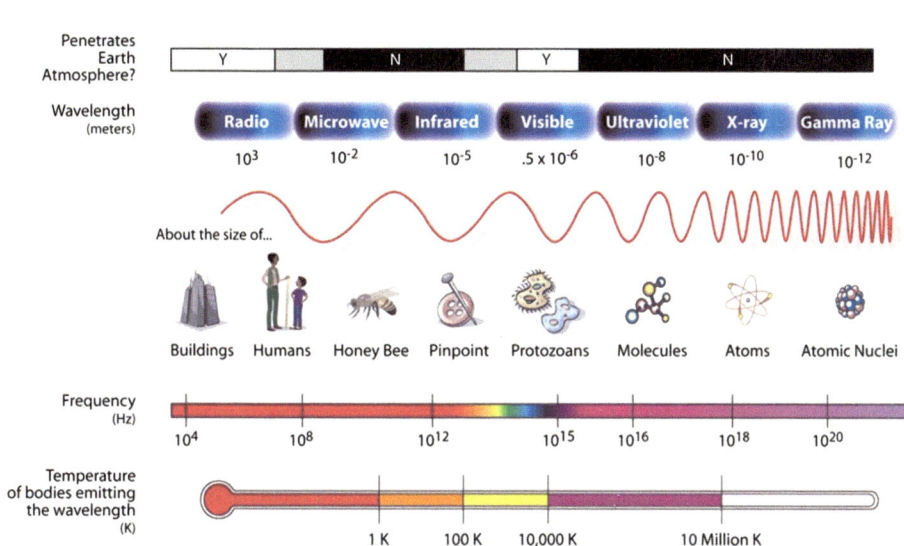

Figure 5.2. The electromagnetic spectrum (Image credit: NASA).

One of the examples where the new quantum approach was hugely successful was Einstein's explanation of the photoelectric effect. The classical approach described light as a wave, but many phenomena, such as the photoelectric effect, proved that light can be considered as particles (photons). This presented a clear paradox – is light a wave or a particle?

This wave-particle duality is an accepted theme in quantum mechanics. Experiments have shown that light could be treated as waves or particles, depending on the nature of the experiment. To be specific, in quantum physics, an elementary particle such as a photon or an electron is both a wave and a particle. It is important to note here that in this new approach the wave or particle is not the most important aspect of existence. This is essentially the idea that has been well explored by authors such as Fritjof Capra. He wrote in his popular work *The Tao of Physics* (2010): "Subatomic particles do not exist but rather show 'tendencies to exist,' and atomic events do not occur with certainty at definite times and in definite ways, but rather show 'tendencies to occur.'

Light (electromagnetic radiation) is one of the basic components of our universe. It is the accelerated movement of an electrical charge that is responsible for the generation of light; in other words, an accelerating charge emits radiation. That part of radiation that is visible to us is called visible light, or simply light.

However, that does not mean visible light is the only electromagnetic radiation in the universe. In fact, visible light is a small fraction of the entire electromagnetic spectrum that we know now as shown in Figure 5.2. The objects in the universe can emit radiation belonging to any part of this electromagnetic spectrum. This is the reason we need telescopes devoted to operate on different wavelengths so that we can ultimately see those objects.

The early history of the universe tells us there was no light in the beginning. The first light formed thousands of years after the beginning of space and time. Light and its speed have a special place in physics. This is one of the foundations on which our entire model of the physical universe is built.

The recent experimental result of the OPERA collaboration 2011 (Oscillation Project with Emulsion-tRacking Apparatus) from the European Organization for Nuclear Research (CERN) has given the notion that the speed of light (c), long considered to be the ultimate speed limit, is vulnerable in terms of maintaining its supremacy in cosmic affairs. This shocking result has had a huge impact. The researchers were baffled, and this result generated immense interest and discussion among the public. Later on, it was found that this result, challenging the speed of light, was erroneous. It is too early to predict the doomsday scenario for the speed of light, as some suggest.

Even with all the attention it gets, there still remain a lot of ambiguity about this well-known constant of nature. So let us begin with the myths associated with speed of light.

FIVE MYTHS ABOUT THE SPEED OF LIGHT

1. *The speed of light is a constant.*
   In fact, this statement is incomplete. The speed of light is a constant as long as it remains in the same material medium. Whenever light enters a denser or lighter medium, its speed and wavelength change, though its frequency stays the same. This explains why the speed of light is slow in water compared to air or a vacuum. The amount of slow down depends on the properties of the material medium, known as the refractive index.

2. *Nothing can travel faster than the speed of light.*
   This is exactly what makes the latest CERN result a popular headline. However, this statement is not as universal as it seems. When one says that no particle can travel faster than the speed of light, what is implied is that a particle with finite rest mass cannot cross the speed of light limit; in other words, it cannot accelerate to a value faster than the speed of light starting from a speed less than the speed of light because that process involves an infinite amount of energy. But particles with no rest mass can travel at the speed of light.

3. *If something travels faster than the speed of light, it goes backward in time.*
   This is the notorious conclusion that keeps the speed of light always on tenterhooks. The special theory of relativity declares that there is time dilation when an object is in motion. The time must run slowly when an object speeds up and remain frozen at the speed of light. Now, beyond the speed of light the relativity equation calculates the time as imaginary; this is open to many different interpretations. It might point to the possibility of going backwards in time, but not a definite conclusion. In order to avoid such paradoxes, Einstein suggested that the cosmic speed limit must be equal to the speed of light. Remarkably, his own theories indicate the probable shortcuts that could be used to travel back in space and time, though they still remain hypothetical.

4. *The speed of light is really fast.*

This is how the ancient Greeks described the nature of light. However, many attempts were made in the past to measure the speed of light. The speed of light is fast but not infinitely fast, as some might think.

The first successful measurement of c was made by the Danish astronomer Olaus Roemer in 1676. Currently, the value of c has been fixed at 299,792.458 km/s. The measurement with atomic clocks and laser interferometers support the precision of this number.

5. *The speed of light is slowing down.*

As perplexing as these mysteries may seem, there is compelling evidence to believe in this speed limit of the universe. Numerous experiments have proved it, and all our scientific understanding, from atomic physics to astronomy, is based on this concept. The speed of light is independent of the motion of the observer, and it does not vary with time or space. However, there are many theories out there that question the unique nature of the speed of light. Some scientists even suggest that it is possible that the speed of light has slowed over billions of years or that it might travel at a different rate at other parts of the universe.

It is reasonable to think that we may have to abandon the cosmic speed limit of light if certain observations or theories become more and more compelling. It may be painful to see the speed of light losing its ultimate speed tag. But that's how science works, abandoning the old in the light of new observations and embracing the novel ideas that can thrust humanity to further limits unknown to us now.

LET THERE BE LIGHT

It's true that implications of the faster-than-light OPERA experiment mentioned earlier are both exhilarating and chilling. As described above, the researchers at CERN presented the astonishing conclusions of their experiments over the last 3 years, which found that particles called neutrinos (the next section will be a detailed discussion on neutrinos) travel at speeds faster than light.

The scientists had been shooting the neutrinos across a distance of 730 km from CERN in Switzerland to the Gran Sasso Laboratory in Italy for the last few years. The primary goal of this experiment was to identify the flavor-flipping nature of these particles, for which they are famously known. There three known different neutrino types, or "flavors," of neutrinos – electron neutrinos, muon neutrinos, and tau neutrinos. For mostly unknown reasons, they can change from one type to the other, a property known as flavor-flipping or the oscillating nature of neutrinos.

However, in the case of the OPERA neutrino experiment, researchers found that the neutrinos arrived at their destination about 60 ns (1 ns is one-billionth of a second) earlier than they should have, had they traveled at the speed of light. The speed of light still remains as one of the foundational pillars of modern science, so the finding has the potential to be exceedingly disruptive. Nevertheless, further investigations proved the fallacy involved in the initial OPERA result, preserving the place of the speed of light.

There are numerous factors to be considered before accepting this rather shocking result. Clearly, as the researchers concluded in the paper, further investigations are necessary to explain the observed effect, which is revolutionary. As one could imagine, it's an extremely difficult experiment and is not a straightforward affair for even a neutrino expert to comprehend the technical details, including the measurements of time and distance involved in the experiment.

The follow-up experiments conducted in this regard have concluded (Antonello 2012) that the neutrinos do not break nature's speed limit. Thus, the whole hype has subsided for now

Neutrinos are neutrally charged with near-zero mass and rarely interact with matter, with billions of them passing through our bodies and trillions passing through Earth every second. They are produced in various nuclear reactions, such as fusion, which powers the Sun. They should be governed by the rules of the special theory of relativity, so it shouldn't be possible to accelerate them to the speed of light, let alone beyond that.

The violation of causality and the possibility of time travel are the two huge implications of traveling beyond the speed of light.

The speed limit of light is the basis of cause and effect: effects always follow causes. Our rational thinking dictates that this order must be followed. We do want to believe, for example, that a bullet leaves the gun before hitting its target. This is obvious in every action and experiment that we observe or perform in nature. But, if the bullet hits the target before it leaves the gun, it would violate causality and would be stupefying. If this were to occur, the basic laws of physics may well have to be rewritten.

Our perception of reality is based on the fact that actions generate outcomes, not the other way around. We shape our destiny on this assumption. But, what if such a lavish free will, long taken for granted, is a mere illusion? Surprisingly, many new studies suggest that our actions are mere execution of a long-ago pre-written script. The cause-and-effect sequence may be pleasing to our superficial minds, but it may have nothing to do with the true underpinnings that make us function in a unique way, which we believe is the result of our own actions.

Also, if particles can travel faster than light and interact with matter, it becomes possible to send information into the past. In other words, time travel into the past would be a possibility. This would arguably destroy our objective sense of past, present, and future that we depend to classify events.

Such a situation will definitely open up many paradoxes that would be very disturbing to live with. Imagine if you receive the information before it was even sent. Hopefully, in the near future, there will be light shed on the topic. Now, let's discuss little more about the neutrinos, or ghost particles as they are popularly known.

GHOST PARTICLES

Enter the world of neutrinos – the improbable invisible particles – and you will not be surprised why they are called ghost particles.

One possible explanation for the neutrinos greater-than-light speed could be that neutrinos are able to access some unknown, hidden dimension of space, which physicists have been hypothesizing for a while. If these particles travel through such shortcuts, they can beat the speed of light even without violating the special theory of relativity. That, too, would open new and exciting vistas in our understanding about the universe, where we are prisoners of the known dimensions in an unknown universe.

How can we not be amused by these tiny particles? They enter everywhere, unnoticed, and strike at the heart of everything, without anyone realizing it. They have been traveling since the birth of the universe. They zip through Earth and continue their passage, a journey that began long before the Solar System was in place. Like ghosts, they are undetectable and disguise their identity. They act like chameleons and morph when we try to catch them. As you read this, billions of them will have entered and exited through your body.

As much as they are abundant, paradoxically, they are hard to detect. It is no surprise that researchers call them the ghost particles. While their better-named cousins, God particles (Higgs Boson), were found just recently, the neutrinos are everywhere. Some researchers even believe these particles account for the missing mass of the universe – oddly termed as dark matter.

Neutrinos are the second most plentiful particles in the universe, just behind photons, the particles that make up light. They come in different flavors, and on their way to Earth they switch their identity, compounding the mystery that surrounds them. In fact, in every second, hundreds of billions of these neutrinos pass through each square inch of our bodies.

Neutrinos are little neutrons, the chargeless particles inside atoms. These tiny neutrons do not reside inside the atoms, like their bigger partner; instead they pass through all material objects as if they weren't present. While neutrons possess mass that makes up the atom, neutrinos were considered massless, at least until recently. These strange characteristics enable them not to interact with anything they encounter, and the consequence is that nothing can catch or trap them. They truly live up to their name – ghost particles.

These particles were introduced to modern physics in 1930 by Wolfgang Pauli, who invented these hypothetical particles in an effort to save the law of conservation of energy, a fundamental law in the universe. The observations of beta decay in radioactive materials demanded that a particle like the neutrino must exist so that this law is not violated.

Neutrinos come from several sources. Most of them were created during the first few moments after the Big Bang. The 14-billion-year journey exhausted them, and they remain mostly undetectable. But, along with microwave radiation, they create the cosmic background radiation, our picture of the infant universe. Other types of neutrinos are produced in exploding stars as well as in laboratory experiments, such as particle accelerators. Cosmic rays, coming from various stellar phenomena that bombard Earth continuously, are also comprised of neutrinos along with other particles.

The Sun is also a source of neutrinos, which are known as solar neutrinos. In the core of our Sun, four protons combine with two electrons to form a helium nucleus, and in that process, two electron neutrinos are created.

A daring experiment spanning several years in a deep South Dakota mine allowed us to observe solar neutrinos. The late Raymond Davis (1964) and his colleagues built a tank of 100,000 gal of cleaning fluid in the depth of the mine to capture the solar neutrinos. Of the billions and billions of these particles coming from the Sun, one would hit an atom of the fluid a day, on average. Raymond Davis was awarded the Nobel Prize for discovering solar neutrinos through this painstaking experiment. Neutrinos, like the ones coming from the Sun, can go through an average 10,000 light-years of lead before interacting once.

Although this was an expected behavior, given the strange properties of these elusive particles, surprisingly a huge disparity existed between the number of neutrinos that arrive on Earth from the Sun and the numbers calculated theoretically that are created in the Sun. In the standard model of fundamental particles, this has come to be known as "the mystery of the missing neutrinos" and has baffled researchers for a long time. The two possible explanations are either that our understanding of the workings of the Sun is incorrect or the theoretical calculations of their numbers are erroneous.

However, a recent experiment has provided an explanation for the "missing neutrinos," although it contributed another dimension to their mystery. Physicists at the Gran Sasso Laboratory in central Italy have been monitoring billions of muon neutrinos beamed to them through Earth from the CERN Laboratory in Geneva, 456 miles away. They had confirmed that a muon neutrino turned into a tau neutrino in an obvious sign of switching identity by a fundamental particle on its way to its destination. Apparently, this could explain why there are fewer numbers of neutrinos arriving on Earth than one expects. This long-sought proof suggests that other types of neutrinos could exist and slip detection. The solar neutrinos on their way to Earth transform to other kinds of neutrinos that we do not yet know.

The personality disorder exhibited by neutrinos that enables them to switch their identity once again underlines the never-ending wonders of the cosmos and the human limitations of comprehending it. One day, neutrinos will be telling us hitherto unknown stories of the universe that the other particles were unable to do. Many theoretical physicists attest to that.

The ability of neutrinos to transform demonstrates that they must have mass, to jump between different flavors, a major deviation from the earlier understanding that these particles are massless. The question of their mass has cosmic importance. For a very long time, it was assumed that neutrinos are massless. However, the new experiments suggest that they could have a tiny bit of mass, maybe even less than that of an electron. Since they are created in vast quantities, the tiny masses could become a significant quantity. Astronomers are unsure what 90 % of the universe is made of, and if the mass of neutrinos can provide part of the answer, then dark matter will not be so dark.

At this point, no such definite conclusion is possible. Researchers are relying on upcoming experiments and detection for clarity. Neutrino detectors are located underground to avoid other particles that could create spurious results. One of the detectors is at the Sudbury Neutrino Observatory (SNO) situated 2 km underground at the Creighton Mine in Ontario, Canada. Physicists think that neutrinos, as demonstrated in the CERN experiment, can switch identities between three known types – electron neutrino, muon neutrino, and tau neutrino – and SNO will be able to detect all three kinds.

Why can't anything go faster than the speed of light (c)? Well, when an object of mass approaches the speed of light, its effective mass increases, as shown by the theory of relativity equations. Consequently, it needs an infinite amount of energy to push the object further. In other words, for an object with mass, it is practically impossible to achieve the speed of light. Even when an object approaches the speed light fairly close the time dilation effects will be significant.

TIME DILATION

Similarly, the special theory of relativity (Einstein 1905) details the time dilation effect when objects move at relatively faster speeds. Put simply, time dilation is the apparent slowing down of time in a fast-moving system when observed by a stationary observer.

The equation for calculating time dilation is as follows:

$$t = \frac{t_0}{\sqrt{1 - \frac{v^2}{c^2}}}$$

where: t=time observed in the other reference frame
t_0=time in observers own frame of reference (rest time)
v=the speed of the moving object
c=the speed of light in a vacuum

As the speed of the object increases the observed time from the other reference frame will appear to be slow. However, in order to observe significant time differences one has to be traveling at really high speeds.

However, there is no such speed restriction on massless particles, as they can travel at the speed of light like the photons that make up light. Though originally assumed to

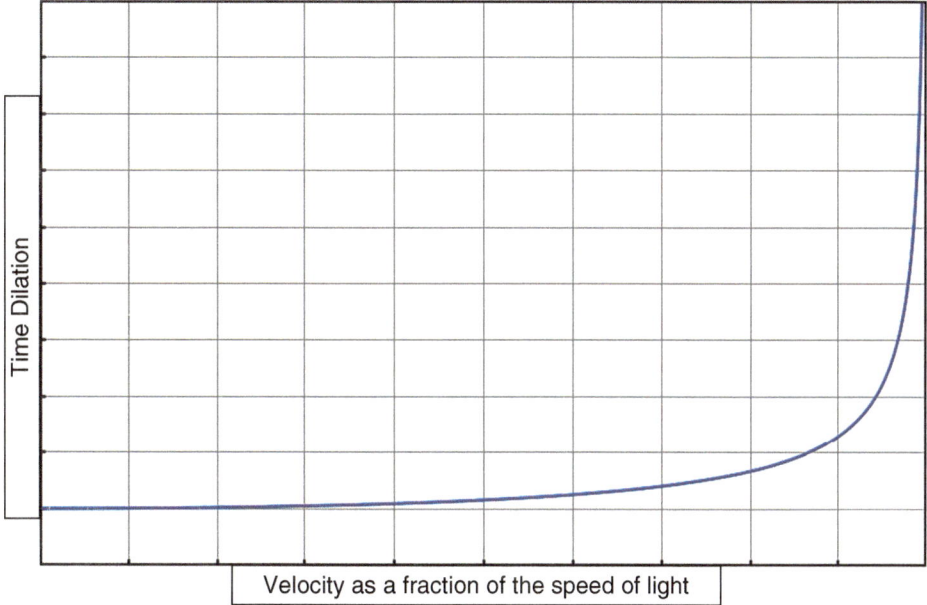

Figure 5.3. The graph showing the effect of time dilation as a function of the speed. The x-axis represents the speed as a fraction of the velocity of light and the y-axis shows time dilation. The x-axis values from 0 to 10 and y-axis values from 0 to 1.

be massless, recent experiments have shown neutrinos possess a small mass and thus they must not have a speed more than that of light. The OPERA result is, seemingly, contradicting this fact. In this experiment, the speed of neutrinos was found to be $1.0000248(28)$ c.

There are several possibilities for real faster-than-light travel that have been proposed, though many of these ideas are hypothetical in nature. In principle, there could be occasions where particles can move faster than the speed of light without violating the theory of relativity.

Although media headlines range from "Einstein may be wrong" to "Time travel may be possible," the speed of light remained as a strong pillar of modern physics. Moreover, some scientists have even invoked exotic extra dimensions to accommodate the result and save the special theory of relativity, when it was first reported.

More reasons to be skeptical is that past studies claiming to have measured particles traveling at speeds greater than c have proved false. One, the 2007 Main Injector Neutrino Oscillation Search (MINOS) experiment in Minnesota, detected neutrinos from the particle-physics facility Fermilab in Illinois arriving slightly ahead of the expected time. However, there was too much uncertainty in the detector's exact position to be sure of the neutrinos' speed and the measurement's significance.

As we know, the meter and the second are arbitrarily defined and are chosen to make precise physical measurements easier. Previously, the meter and second have been defined in various ways according to the measurement techniques of the time, and they could change again in the future. The latest experimental result is not significant enough to withstand the possible small change that could be reflected in the accepted value of the speed of light with a new definition of meter and second.

The OPERA scientists asked for the replication of the experiment before any exotic conclusions be drawn. As the proverb goes, "It's better to light a candle than curse the darkness."

Playing with light is fun and rewarding, yet it is not time to write it off as the speed limit imposed on the universe.

As discussed above time dilation is an inescapable consequence of the special theory of relativity, which imposed a speed limit on the speed of light. The time dilation can be precisely calculated using the time dilation equation that you have seen earlier.

Have humans ever imagined different planes of time? Probably, we can find such thoughts in ancient mythology. A story that narrates the experience of a king and his daughter is part of the Bhagavata Purana, an ancient Hindu text.

The story goes like this. Kakudmi was a generous king, and he had a very beautiful daughter named Revati, gifted with all auspicious qualities. The king had mysterious powers that enabled him to take his daughter Revati with him and travel to Brahma-loka[1] and to speak with Brahma whenever he desired. The king wanted to seek his advice about a proper husband for his daughter.

Upon their arrival, they noticed Brahma was attending a musical performance by the Gandharvas.[2] The king and his daughter waited patiently until the performance was over.

---

[1] The plane of existence where Brahma resides.

[2] Gandharvas are heavenly beings and often act as messengers between Gods and humans.

Then Kakudmi made his request and presented the list of prospective husband candidates for his daughter. Brahma explained that time runs differently on different planes of existence, and that during the short time they had waited in Brahma-loka to see him, many cycles of time had passed on Earth.

To King Kakudmi's astonishment Brahma said to him, "O King, all those whom you may have decided within the core of your heart to accept as your son-in-law have died in the course of time. Those upon whom you may have already decided are now gone, and so are their sons, grandsons and other descendants. You must therefore bestow your daughter upon some other husband, for you are now alone, and your friends, your ministers, servants, wives, kinsmen, armies, and treasures, have long since been swept away by the hand of time."

To the disappointed king, Brahma offered a solution primarily to comfort him. Brahma explained that Vishnu, the preserver, was currently incarnated on Earth in the forms of Balarama,[3] and he was worthy husband for Revati.

The king and his daughter returned to Earth and were amazed by the changes that had taken place while they were away from Earth. They found an earth evolved in several dimensions. They found Balarama, as advised by Brahma, and proposed the marriage, which was accepted. The marriage was then duly celebrated.

This story resembles the twin paradox in a commonly used sketch to derive the time dilation effect, in which a twin left on Earth would age slowly compared to the other one who went for a trip at relativistic speed for years. Our current technology can't accomplish relativistic speed travel, but the minimal effect of time dilation can be felt even when astronauts travel a considerable amount of time. An astronaut on the ISS would have aged slightly less than everyone else on the ground because of this effect. The calculations show that this only about 0.007 of a second because the speed involved in this case is not much. However, time dilation is a fact, and we will experience the effects significantly when relativistic speeds become the norm.

THE CONSTANTS IN NATURE

In the beginning of this chapter we briefly discussed the concept of universal constants, especially the specific case of the speed of light. Now let's discuss this topic in a general sense. A refreshing approach to this subject has been introduced in recent years.

What are the constants of nature? Where do they come from? How do we know they are truly constants? These questions are much deeper than we think and have been discussed in scientific circles for a long time.

It has been universally accepted that there are three fundamental physical constants in the universe. These are the universal gravitational constant (G), the speed of light (c) and Planck's constant (h). Mathematical equations or physics theories all employ these constants. However, no one has ever explained any of the constants or why constants take special numerical values.

---

[3] Balarama is the ninth avatar (incarnation) of Vishnu and is the elder brother of Krishna.

Scientists look for rhythms and patterns in the nature and then they model that behavior, which later on become the laws of nature. Once such an ideal becomes a law, we should be able to make predictions based on it, and it also should be able to explain experimental observations. The constants arise when scientists formulate the mathematical expressions to model the behavior of nature. They assume that the numbers that represent these constants remain unchanged throughout the universe. In other words, at any part of the universe these numbers must remain the same.

For example, consider Newton's formula for the force of gravity.

$$F = \frac{GMm}{r^2}$$

where G is the "universal gravitational constant" and is 6.67384×10^{-11} m^3 kg^{-1} s^{-2}.

The symbols in the equation masses (M and m) of the objects and the distance (r) between them can vary.

Whether we calculate the force of attraction between Earth and the Moon or Earth and the Sun the above equation will employ the same G value with varying masses and distances. So when Newton formulated this equation he knew that the force between two objects is proportional to the product of their masses and inversely proportional to the square of the distance between them. Since this relation holds true for any pair of objects the proportion was replaced by a constant G. Newton introduced these constants in his equation for the first time in 1687. In 1798 the English physicist Henry Cavendish measured the very tiny force between two lead masses by using a very sensitive torsion balance and so determined the above mentioned numerical value for G.

In a very similar fashion, the special theory of relativity established the speed of light (c) as a constant, and as mentioned in the beginning of the chapter Planck introduced the Planck's constant (h).

Again in the 1900s, Max Planck reasoned that his newly discovered fundamental constant (h) could be combined with the other known and apparently universal constants G and c to form a unique set of mass, length and time parameters. He argued that the other observers in the universe would introduce the same units regardless of their location, if these units were based on fundamental constants.

These parameters, namely, the Planck mass (M), the Planck length (R) and the Planck time (T) are derived using the fundamental constants, and they constitute Planck scale. These parameters derived by Planck may not have any normal applications, yet they remain as the absolute limits beyond which we can't even have our theories applied.

For example, the calculated value of Planck length is 1.62×10^{-35} m, which is about 10^{20} smaller than the size of a proton. At this scale the quantum effect of gravity becomes strong, and the smooth structure of space-time breaks down into some as yet unknown structures. Since we do not yet have a dependable framework to combine general relativity and quantum mechanics, the laws of physics may break down on the Planck scale.

It is interesting to note that we are not absolutely certain about the laws of physics and the constants, though in physics we assume that these constants, as their name indicate, remain unchanged. No wonder why some physicists have expressed some cautious skepticism about these constants even in the past.

In a lecture entitled "The Relation of Physics to Other Sciences," (1963) Richard Feynman observed:

> Here are the laws of physics, how did they get that way? We do not imagine, at the moment, that the laws of physics are somehow changing with time that they were different in the past than they are at present. Of course, they may be, and the moment we find they are, the historical question of physics will be wrapped up with the rest of the history of the universe, and then the physicist will be talking about the same problems as astronomers, geologists, and biologists.

Moreover, since the time of Feynman, some observations have forced this once unthinkable view into debate. A team of astronomers reported (Webb et al. 2011) that they have observational evidence of variation in the fine structure constant.[4] If true, that observation would overturn scientists' basic assumption that the laws of physics are the same everywhere in the universe.

Additionally, the implications of these constants are much more than what we would imagine. Some argue that these constants are fine-tuned so that the universe works the way it is. The anthropic principle, which claims that the universe appears to have been fine-tuned for our own existence, is often at the center of controversy. If any of the fundamental constants has a different numerical value, the universe as we know should not be there.

There are reasons to accept a weak form of the anthropic principle, in as much as the building blocks of the universe require such an environment. However, the strong anthropic principle as advocated by some is beyond the limits of our rational science. Some of the harsh criticism is directed towards the so-called strong anthropic principle. As physicist and author Victor J. Stenger (1985) remarked in the *New Encyclopedia of Unbelief*, "In short, much of the so-called fine-tuning of the parameters of microphysics is in the eye of the beholder, not always sufficiently versed in physics, who plays with the numbers until they seem to support a prior belief that was based on something other than objective scientific analysis."

Although there are reasons to be skeptical about any intentional fine tuning argumenta, it is equally important to recognize how science depends on the constants in nature to make sense of the universe. Our scientific foundation is based on the premise that the laws of physics and the constants of nature must remain immutable irrespective of space time constraints.

The constants appear in mathematics as well. Here we will share the story of the most popular constant in human history – pi.

LIFE OF PI

The fascination with pi, the most famous number in human history, is as limitless as its own digits. Thousands of years ago, pi was born out of the pure desire to perfect the measurement of geometrical shapes, such as circles. However, since its birth pi has evolved to an entity, more than just a number, which is often linked to mysticism.

---

[4] The fine-structure constant is equal to 1/137. It measures the strength of the electromagnetic force that governs how electrically charged elementary particles and light (photons) interact.

However, as pi celebrates another day, some mathematicians are calling for its demise – sending shock waves through the pi community.

Pi day, observed on March 14, was founded by physicist Larry Shaw, a tradition that began at the San Francisco Exploratorium. Not surprisingly, 3/14 represents the first three digits of the never ending digits of pi (3.14159....). In the shadow of these celebrations, many forget the fact that March 14 is known for another reason – Albert Einstein's birthday.

The relation between the circumference and diameter of a circle has been known for thousands of years. The ancient Babylonians calculated the area of a circle by multiplying the square of its radius by three, indicating a value close to three for pi. Nonetheless, it was the ancient Greek mathematician Archimedes of Syracuse who calculated the value of pi somewhat accurately. He concluded an average value of 3.1485 for pi – somewhat close to the current known value. Yet the symbol π, to represent this important ratio between circumference and diameter of any circle, appeared in mathematics a great deal later. Some approximations, such as 22/7 and 355/113, had also been used to express this ratio.

Pi being an irrational number, the above fractional representations are mere approximations, and the endless digits that make up pi do not show any patterns. Pi is also transcendental, as it cannot be the solution of an algebraic equation of any degree with rational number coefficients.

A relatively unknown mathematician named William Jones is credited with using the symbol for the first time around 1706. However, the symbol got universal acceptance when the great Swiss-born mathematician Leonhard Euler put π into much wider use in mathematics. Since then, more and more digits of pi have been calculated, with the current world record reaching 10 trillion digits. The irrational pi will continue to leave a trail of digits with no end in sight.

Why would someone care about the long list of random numbers in pi, knowing that they will continue infinitely? The obsession with pi is more mysterious than pi itself. There are events ranging from pi procession to the annual pi party on March 14. Some recite pi digits to claim the best record and others write in pilish or even produce pi-poems known as "piems."

Simply put, some people are just crazy about pi. Some see beauty and uniqueness in the non-ending digits of pi. The idea of writing successive words in a sentence with a length that represents the digits of the number π (= 3.14159265358979...) is termed as pilish. An example of such a writing, believed to have been composed by the English physicist Sir James Jeans: How I need a drink, alcoholic in nature, after the heavy lectures involving quantum mechanics!

(3. 1 4 1 5 9 2 6 5 3 5 8 9 7 9)

In 2009, the U. S. Congress voted to officially recognize March 14 as Pi Day to encourage schools and educators to observe the day with appropriate activities that teach students about pi and engage them in the study of mathematics. Although the vote was not quite unanimous, luckily there was no filibuster or sequester on the proposal.

Whereas the devotion to pi is displayed and may feel like religious zeal for some, there are questions raised about its existence.

In an article entitled "Pi is Wrong!", mathematician Bob Palais argued that the true circle constant must be the ratio that represents circumference and radius. In other words, pi must be replaced by 2 pi, which would make a full circle to turn pi radians instead of the current 2 pi radians.

$$2\,pi = circumference\,/\,radius$$

The idea of replacing pi with the Greek letter tau (τ) that represents 2pi gained some momentum when Michael Hartl, physicist and mathematician and author of "The Tau Manifesto," claimed many advantages of using tau over pi. However, pi remains as an object embedded so strongly in our popular culture that it is not ready to retreat anytime soon.

A NEW PANORAMA

What do all these discussions point to? Are we poised for a new panorama? Are there undiscovered particles and laws of nature? In the quest to understand the fundamental nature of the universe, physicists have modeled a universe with the so-called elementary particles that make up everything. But as they explore more closely, the secret life of fundamental particles is unveiling a nature of reality beyond our imagination.

Although the mysteries about these particles persist, some researchers are considering the practical applications of neutrinos. As an example, underwater communication, such as for submarines, is in its infancy even now. Radio waves, our most common tool of communication, do not travel well in water, restricting underwater communication rates to a few bytes per second. Since neutrinos can travel through any material, including water, almost at the speed of light, researchers see a potential in new forms of underwater communication using these particles. The ghost particles might 1 day become a tool of communication considered hitherto impossible.

The enthusiasm about neutrino research is so widespread that many researchers are joining the particle hunt adventure. These most un-interactive particles can liberate our imaginations and are capable of revealing the hidden places in the cosmos. Star explosions, such as supernovae, emit a huge amount of neutrinos, providing the opportunity for studying the core of stars and their inner dynamics. As these particles pass through intervening matter, they can provide us with a magic vision of the cosmos, if we can catch them.

Among the numerous particles that make up our universe, neutrinos are the tiniest and mostly sterile, yet their implications are profound. A coherent theory of fundamental reality has to account for the mysteries of these ghost particles.

Playing with light might be one of the ways in which we can better understand nature. When evidences pile up, we may have to abolish some of our long held notions and laws. In his *Scientific Autobiography and Other Papers*, Planck remarked that "A new scientific truth does not triumph by convincing its opponents and making them see the light, but rather because its opponents eventually die and a new generation grows up that is familiar with it."

REFERENCES

Alexander Pope. (2008). *The works of Alexander Pope* (p. 172). Edition. Bibliolife. Charleston.

Amati, D., Ciafaloni, M., & Veneziano, G. (1989). Can spacetime be probed below the string size? *Physics Letters B, 216*(1–2), 41–47.

Antonello, M. (2012). Measurement of the neutrino velocity with the ICARUS detector at the CNGS beam. *Arxiv.org. V1*, 1–13. Available at: http://arxiv.org/abs/1203.3433. Accessed 12 Sept 2012.

Barrow, J. D. (2005). Varying constants. *Philosophical Trans-Actions of the Royal Society London, A363*, 2139–2153.

Barrow, J. D., & Frank, J. T. (1986). *The anthropic cosmological principle*. Oxford: Oxford University Press. 21.8.

Barut, A. O. (1988). *Foundations of Physics, 18*(1), 95–105.

Bassuk, D. E. (1987). *Incarnation in Hinduism and Christianity: the myth of the god-man*. Atlantic Highlands: Humanities Press International.

Davies, P., Davis, T., & Lineweaver, C. (2002). Black holes constrain cosmological constants. *Nature, 418*, 602.

Davis, R., Jr. (1964). *Physical Review Letters, 12*, 303.

Dirac, P. A. M. (1937). The cosmological constants. *Nature, 139*, 323.

Eddington, A. S. (1923). *The mathematical theory of relativity*. London: Cambridge University press.

Einstein, A. (1905). On the electrodynamics of moving body. *Annals of Physics, 322*(17), 891–921. Translation into English: Lorentz, H., Einstein, A., & Minkowsky, H. (1923). *The principle of relativity*. London: Methuen, p. 35.

Einstein, A. (1922). *The meaning of relativity 1974* (5th ed.). Princeton: Princeton University Press.

Feather, N. (1959). *Mass, length and time*. Edinburgh: University Press.

Feynman, R. P. (1995). *Six easy pieces (Chapter 3)*. Reading: Addison-Wesley. ISBN 0-201-40896-1.

Feynman, R. P., Leighton, R. B., & Sands, M. (1963). *The Feynman lectures on physics vol. 1 (Chapter 3)*. Reading: Addison-Wesley.

Flynn, T. (Ed.). (1985). *The encyclopedia of unbelief, volumes I and II* (1st ed.). Amherst: Prometheus Books.

Fritjof, C. (2010). *The Tao of physics: An exploration of the parallels between modern physics and eastern mysticism* (5th ed.). Boston: Shambhala.

Fritzsch, H. (2009). *The fundamental constants: A mystery of physics*. Hackensack: World Scientific Publishing Company. Edition.

Fundamental Constant May Depend on Where in the Universe You Are – sciencenow. (2013). Fundamental constant may depend on where in the universe you are – Sciencenow. Available at: http://news.sciencemag.org/sciencenow/2011/11/fundamental-constant-may-depend-.html. Accessed 05 June 2013.

Garay, L. G. (1995). Quantum gravity and minimum length. *International journal of modern physics. A, Particles and fields, gravitation, cosmology, A10*, 145–165.

Grossman, L. (2011). Faster-than-light neutrinos? New answers flood in – Physics-math – 05 October 2011 – new scientist. *Science news and science jobs from New Scientist – New Scientist*. http://www.newscientist.com/article/dn21010-fasterthanlight-neutrinos-new-answers-flood-in.html. Accessed 12 Oct 2011.

Heilbron, J. L. (1986). *Dilemmas of an upright man: Max Planck and the fortunes of German science* (1st ed.). Cambridge: Harvard University Press.

Kalinski, M., et al. (2003). *Physical Review A, 67*, 032503.

Kantha, L. (2012). What if the gravitational constant G is not a true constant? *Physics Essays, 25*(2), 282–289. Academic Search Premier, ebscohost, viewed 6.

Krauss, L. M. (1998). The end of the age problem and the case for a cosmological constant revisited. *Astrophysical Journal, 501*, 461–466.

Krauss, L. M., & Starkman, G. D. (2000). Life, the universe, and nothing: Life and death in an ever-expanding universe. *Astrophysical Journal, 531*, 22–30.

Loeb, A., & Steven, R. F. (2013). *The first galaxies in the universe*. Princeton: Princeton University Press.

Loeb, A., Ferrara, A., & Richard, S. E. (2008). *First light in the universe*. Berlin: Springer.

Max Planck – Biography. (2012). *Max Planck – Biography*. Available at: http://www.nobelprize.org/nobel_prizes/physics/laureates/1918/planck-bio.html. Accessed 04 June 2012.

Okun, L.(1991). The fundamental constants of physics. Soviet Physics Uspekhi, *3499,* 177.

Pi Day. (2013). Events, activities, & history|exploratorium. 2013. Pi Day 2013: Events, activities, & history|exploratorium. Available at: http://www.exploratorium.edu/pi/. Accessed 12 Mar 2013.

Planck, M. (1968). *Scientific autobiography and other papers*. New York: Philosophical Library. Edition.

Rajagopalachari, C. (1967). *Bhagavad-Gita* (3rd ed.). Bombay: Bharatiya Vidya Bhavan.

Smeulders, P. (2012). Why the speed of light is not a constant. *Journal of Modern Physics, 3*(4), 345–349. Academic Search Premier, ebscohost, viewed 6.

Smolin, L. (1997). *The life of the cosmos* (1st ed.). New York: Oxford University Press.

Stenger, V. J. (1995). *The unconscious quantum: Metaphysics in modern physics and cosmology*. Amherst: Prometheus Books.

Veneziano, G. (1986). A stringy nature needs just two constants. *Europhysics Letters, 2*, 199–204.

What's Light. (1999). What's Light. [ONLINE] Available at: http://home.fnal.gov/~pompos/light/light_page1.html. [Accessed 12 October 2013].

Webb, J. K., et al. (2011). Indications of a spatial variation of the fine structure constant. *Physical Review Letters, 107*(19), 5.

6

Dark Forces in the Universe

"Equipped with his five senses, man explores the universe around him and calls the adventure Science."

–Edwin Hubble 1954

We often say "it makes sense," which implies that it is possible to experience *it*, whatever that be, through our senses. Should we insist or enforce that condition on everything in the universe? We can argue for that; however, the universe does not need to gratify our senses. Even when we can't experience something, we can reason certain things and indirectly experience the presence of the unknown. That does not diminish the nature of scientific wisdom but, in fact, accelerates our understanding of the big picture.

Dark energy and dark matter together contribute to almost 96% of the total energy-mass of the universe. They have the potential to tear up our whole picture of the universe that we put together. The very nature of each seems to be completely different from the other, and so they need to be discussed separately.

DARK ENERGY

Almost three-quarters of our universe is comprised of dark energy, the newest bewildering component of our cosmos, so believe astronomers. To further exacerbate the situation, we know very little about it. We live in a universe ruled by the unknown.

The discovery of the accelerating nature of the universe's expansion demands a force surpassing gravity and stretching the fabric of the cosmos. Physicists like to call this dark energy, and the Big Bang model predicts that the universe consists of 73% dark energy and 23% dark matter. Normal matter, which constitutes just 4% of the universe, makes up all the stars, planets, and living and non-living things that are known to us.

Scientists estimate the changes in expansion rate by comparing the redshifts of distant galaxies with the apparent brightness of Type 1a supernovae found in them (which will be

S. Mathew, *Essays on the Frontiers of Modern Astrophysics and Cosmology*, Springer Praxis Books, DOI 10.1007/978-3-319-01887-4_6, © Springer International Publishing Switzerland 2014

discussed later). These measurements suggest that the expansion of the universe is accelerating. One such study was conducted by Adam Riess (1998) of the Space Telescope Institute and John Hopkins University. Riess used Hubble Space Telescope data to improve the value of the expansion rate of the universe (the Hubble constant) to an accuracy of 3%. Ironically, this implies that the dark energy, as Albert Einstein assumed, is steadily pushing the fabric of the universe. Riess and other researchers would eventually like to see the Hubble constant refined to a value with an error of no more than 1%, to put even tighter constraints on solutions to the dark energy puzzle. We should be optimistic of the fact that science is closing in to comprehend one of the most baffling concepts of modern day astronomy.

The Big Bang model needs dark energy to account for the current state of the universe, although scientists have no lucid explanation for it. In fact, some scientists believe dark energy is the same as the cosmological constant originally introduced by Einstein to counterbalance the gravitational attraction of matter in the universe. He later discarded the cosmological constant when Hubble's observations proved the expansion of the universe. Now, modern cosmologists think Einstein's cosmological constant may hold the key to the accelerated expansion of the universe.

Until just a decade ago, scientists were near unanimous in their assumptions on the fate of the universe. They thought that the cosmic expansion triggered by the Big Bang would slow down over time or at least it would remain unchanged. Gravity, being an

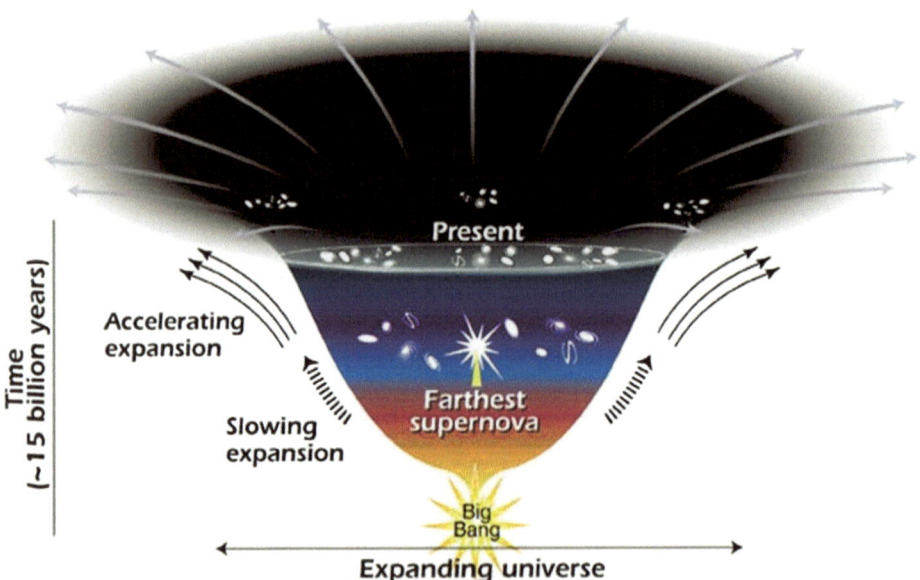

Figure 6.1. Living in an expanding universe. The observation shows that the universe is not only expanding but that the expansion is accelerating (Image credit: NASA).

attractive long-range force, pulls the components of the universe together, thereby prohibiting unruly expansion. So they set out to assess the expansion speed of the universe. Contrary to their expectations, they found a universe that is accelerating (Adam et al. 1998). The data they collected were contrary to their expectations and the long held notion on the nature of our universe.

The evidence of the unexpected expansion of the universe came from observations of distant supernovae explosions. These are the brightest explosions that can be seen over vast distances across the universe. Telescopes measure the fine prints of such explosions.

When light begins its long journey from an exploding supernova, it has to confront an expanding universe along its path. This expansion of space redshifts the light waves. In other words, the wavelength of light is stretched to the red end of the spectrum. The level of redshift and the distance of the explosion allow us to measure the rate of the expansion of the universe. They record the expansion history of the universe.

The observations are convincing enough to conclude that some unknown force is prompting the acceleration. This force is greater than that neutralizing the combined gravitational pull of normal and dark matter.

This discovery wasn't cause for celebration among astronomers – rather, it was shocking. It portrays a world vastly different than the one we are used to. We expect a tossed object to fall back to Earth unless a force overcomes the gravity of our planet. On a grand scale, such an unknown force is spurring the growth of the universe by flinging galaxies apart. To accommodate that large repulsive force accelerating the universe, scientists coined the term dark energy.

Einstein's theory of relativity hinted at a repulsive force in the universe. He incorporated the idea of a cosmological constant to support the idea of a static universe. He reasoned that for the universe to remain static, this repulsive nature of the cosmological constant would surpass the effect of gravity. When Edwin Hubble's observation of galaxies proved the expansion of the universe as opposed to a static universe, Einstein retracted his original idea, calling it the "biggest blunder" of his career. Now, decades after his death, some cosmologists think that Einstein's blunder may come to the rescue. They identify the cosmological constant as one of candidates for explaining dark energy.

The most recent observations by the WMAP satellite provide evidence of dark energy in the universe. In the absence of dark energy, the mass-energy density of the universe is not sufficient to explain the flat geometrical shape of the observable universe.

What if these observations are fooling us? Are we missing some fundamental truth about the universe? After all, we infer the existence of dark energy only from the accelerating universe. We expect the universe to act according to our known laws. Modifying Newtonian and Einsteinian laws would have a startling impact on how gravity works. It may eliminate the need for dark energy. But astronomers are not prepared to admit the demise of dark energy yet. A series of new tests using Earth- and space-based telescopes are planned to identify the ghostly energy.

Additionally, quantum mechanics views the cosmological constant as the "energy of the vacuum," or the energy of empty space. It is assumed that all space is filled with this form of energy. This repulsive background energy associated with the empty space could be dark energy. However, there is no compelling evidence for that claim. The theoretical calculations suggest that the amount of vacuum energy is too high for reasonable explanations. Vacuum energy is the most straightforward explanation for dark energy, which is equivalent to the cosmological constant. So far, the data from the various probes of dark energy are consistent with a constant value for the vacuum energy.

To understand vacuum energy we have to abandon our conventional wisdom on empty space. The empty space, or the vacuum, is not truly as empty as we would like to imagine. The Heisenberg uncertainty principle allowed for particles to blink into and out of existence on extremely short times scales. This is because, when there is greater uncertainty in energy, the uncertainty related to time is smaller. In other words, we are uncertain about the existence of particles, and for a very short time these particles exist without being noticed – virtual particles. The vacuum energy may be the result of these virtual particles.

However, there remains a huge discrepancy between the observed and estimated vacuum energy density required to explain dark energy. When physicists tried to estimate the energy density associated with the quantum vacuum, it led to absurd results. The vacuum energy they calculated was unimaginably larger than what is observed. To be specific, the number came out 10^{120} times more than the required value. That is too big of a mistake.

Based on such arguments, we could agree upon the fact that emptiness is not a true void, as it was deemed in the past. Quantum theory considers a vacuum as a pool of virtual particles rapidly popping in and out of existence. The particles and energy inside the so-called emptiness is a result of invisible interactions. Other explanations point to the possibility of extra dimensions. The dark energy might be lurking in those dimensions while we are stuck in our three-dimensional world. While it is imprisoned in those dimensions, its gravitational effects can reach out into our space. Its gravitational force may be repulsive in our universe.

Another possible candidate for dark energy is "quintessence," a new kind of dynamical energy or field that fills all of space. In a paper entitled "A quintessential introduction to dark energy," physicist Paul Steinhardt describes quintessence as a dynamical, evolving, spatially inhomogeneous component with negative pressure (2003). This has an opposite effect on the expansion of the universe, like negative pressure, compared to that of matter and normal energy. Some theorists have named this after the fifth element of the Greek philosophers.

Even with all these different explanations, the mystery about dark energy continues, as we still don't know what it is like, what it interacts with, or why it exists.

Dark energy seems to play a huge role in determining the fate of the universe. It could cause the universe to violently expand ("Big Rip") or make the universe implode ("Big Crunch").

The concept of five classical elements has its roots in many ancient cultures, including Babylonian mythology. In ancient Greek, quintessence means "after quint," or fifth. The fifth element was referred to as ether, the material that fills the universe. A similar concept is described in ancient Hindu texts.

In Hinduism, the four states of matter are air, fire, earth, and water. The fifth element described as Akasa (sky or void) represents the one beyond the material world. Every material creation is due these five elements, and upon destruction of a material object, it dissolves into this element. The fifth element Akasa is like Brahman – formless, limitless, and imperishable. By its very nature, Akasa is beyond the senses and incomprehensible. Interestingly, these thoughts are prevalent in many of the ancient cultural systems.

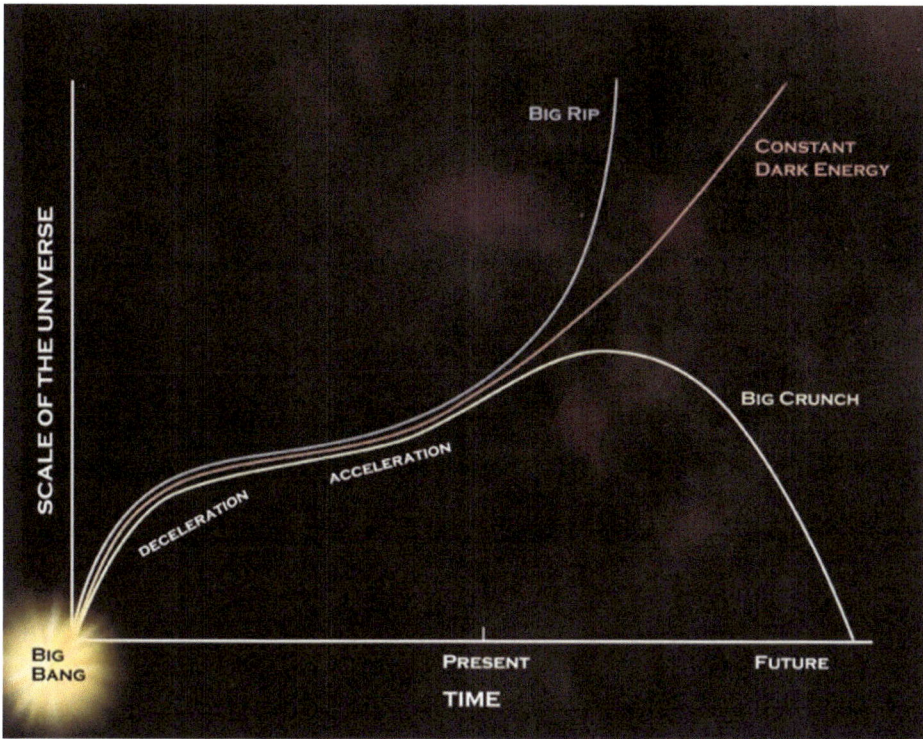

Figure 6.2. Dark energy may dictate the fate of the universe (Image courtesy of NASA/STScI/Ann Feild. Used with permission).

Similarly, Tibetan Buddhism presents the concept of impermanence, which asserts that nothing is free from decay. Similar to the way the individual sentient being is born, dies, and is reborn, so does the universe. As in Hinduism, any cosmological system is seen to come into being, exist for billions of years, and then dissolve, before coming into existence again. However, Buddhists did not believe in the existence of a permanent and fixed reality that could be referred to as Brahman. There is the destruction of one system, in which the fire, earth, air, and water particles separate and conceivably fall apart, but space or emptiness remain.

By analogy, in modern cosmology, quintessence would be the constituent of the universe, unknown to us, at least for now, beyond the other known ones such as particles and energy. Again, in his paper, "A quintessential introduction to dark energy," physicist Steinhardt (2003) writes that not only the immediate future of the universe will be governed by dark energy, but it will play a more profound role in the history of the universe. Dark energy will determine the rate of dilution and cooling of the matter and energy and thereby the fate of the universe in long-term. Similarly, dark energy could explain the cyclic nature of the universe, which was described in the previous chapter. Dark energy could be the key part of the engine driving the periodic evolution of the universe.

Finally, some researchers suggest that we need to take a fresh look at the law of gravity, which might need amendments or even the revamping of the whole theory itself. Even Einstein recognized that his theory of general relativity is not a perfect one. To explain the universe in its entirety, he tried to reconcile his theory with quantum mechanics without success. In fact, he spent the last three decades of life towards this goal.

In a paper entitled "Constraining inverse curvature gravity with supernovae," Olga et al. (2005) argued that the current accelerated expansion of the universe can be explained without resorting to dark energy.

We have to admit that all these are scientific speculations until supporting evidence is lined up. Unfortunately, our observations and measurements depend upon something we cannot comprehend yet. We know the laws of gravity account for the behavior of our Solar System pretty well. But, how do we know if our laws of gravity break down over huge distances? Or even worse, it is possible that we are unaware of some long-range forces other than gravity.

Whether dark energy is a property of space, a new dynamic fluid, or a new theory of gravity could be resolved with more accurate data. Determined to break the impasse, NASA and the U. S. Department of Energy have together decided to develop a space-based dark energy investigation. This strategic mission, known as JDEM – Joint Dark Energy Mission – is currently in the design phase. The JDEM observatory is set to launch in 2016 to study the properties of dark energy and assess the expansion rate of the universe over different time periods during its cosmic evolution.

Beginning with Newton, physics focused on a universe filled with matter and radiation, and most of our theories reasonably explains the interplay of these two. Now, we know that these two components make up only small part of the universe. The self-repulsive nature of dark energy is completely new to physics, which has been so used to the attractive nature of gravity.

Until we grasp this dynamism categorically, we have to confront the dark forces of nature. If this phantom energy exists, it will rip our universe apart, leaving us alone in the Milky Way surrounded by huge emptiness. If it doesn't, for our future generations, dark energy may come to symbolize the extent of our ignorance.

SHINING A LIGHT ON DARK MATTER

What we see constitutes just a small part of our universe. We know nothing of the remaining, except that it exercises gravitational pull on observable objects. Where should we turn for clarifications? Surprisingly, some scientists go underground.

Inside an old iron mine in northern Minnesota, an array of cryogenic detectors are patiently attempting to capture the fingerprints of dark matter, which makes up 90% of the entire matter in the universe. In December, 2011 they reported (Ahmed et al. 2011) a possible whiff of dark matter. If dark matter truly exists, it would rewrite our cosmic ancestry, for sure.

We know that the elementary particles, such as protons and electrons, compose our visible universe. Anything from stars to planets to single-cell organisms are all the products of these particles, which are known as baryonic matter.

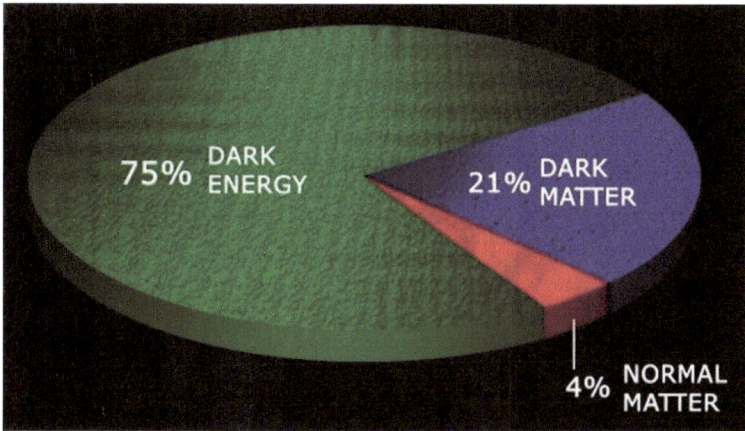

Figure 6.3. The universe is dark. Dark energy is estimated to contribute about 73% of the universe, dark matter about 23%, and normal matter about 4% (Image credit: NASA/Courtesy of nasaimages.org).

We are tempted to think that the whole universe is filled with baryonic matter in different forms – humans, animals, planets, stars, and galaxies. But cosmic researchers believe that this type of matter accounts for only a small fraction of all matter in the universe – just 10%. The detection of this ordinary form of matter, as we know it, is possible through types of electromagnetic radiation ranging from gamma rays to radio waves. On the other hand, the presence of dark matter, which is assumed to make up 90% of the universe, is known only through the gravitational forces it exerts on other objects.

Scientists believe that dark matter is made up of non-baryonic exotic particles, such as WIMPS, or Weakly Interacting Massive Particles. The interaction of WIMPS with ordinary matter is a delicate phenomenon. Sporadically, these particles scatter from the normal atoms. Under ideal conditions, such scattering can be spotted as it produces detectable traces of energy. That's what the scientists are attempting in Minnesota's Soudan Underground Mine, which houses the Cryogenic Dark Matter Search (CDMS) experiment, some 800 m underground.

The supporting evidence for dark matter comes from the study of galaxies and clusters. In 1960s, astronomers and theoretical physicists were met with a new mystery that still remains unsolved – dubbed the "missing mass problem." This was essentially the result of the extensive studies they carried out on the internal motions of galaxies within the clusters. Apparently clusters are groups of galaxies bound together by the force of gravity similar to that of star clusters. Most galaxies are part of clusters that contain a handful to hundreds of galaxies. Ideally, the motion of individual galaxies could be explained on the basis of the laws of gravity. In other words, when there is more mass in the cluster the individual galaxies must move faster. However, the observations revealed that such rules are not followed in most clusters.

It is possible to deduce the mass of the cluster using the laws of gravity. This is similar to determining the mass of the Sun – the center of the Solar System – by knowing the

Figure 6.4. The invisible universe. The dark matter constitutes most of our universe, and it does not interact with the electromagnetic force – the invisible universe (Image credit: NASA).

velocity and distance of the planets from the Sun. Now, it seems that the laws of gravity must be equally valid in the realm of planets, stars, or galaxies. The measured velocities of the galaxies indicate that there must be much more mass in the cluster than the visible matter. That's where dark matter emerges. Even with modern-day observations, this discrepancy between visible matter and the observed motion remains.

Astronomers use pure Newtonian physics to find the weight of a spiral galaxy. They can measure the rotational velocity v of gas clouds orbiting a distance r from the galaxy.

From the known values of v and radius r the centripetal acceleration of clouds$= v^2/r$. Equate acceleration to gravitational pull of the matter M inside the orbit$=GM/r^2$ (G is the universal gravitational constant$=6.67\times10^{-11}$). Finally, solve the equation for $M = rv^2/G$.

In fact, the velocity of the cluster galaxies was found to be more than could be accounted for by the individual galaxy masses. The conclusion is that some unknown form of mass is contributing to this anomaly, which is much more than the known masses of all the galaxies together. The actual measurement and calculations showed that about 95% of the mass is missing, which implies that normal matter is rare in the universe.

There is another layer of evidence that can be linked to the existence of dark matter, which arises not from the galaxy clusters as explained above but from the individual galaxies themselves.

Even in our own Milky Way Galaxy, about 200 billion stars orbit around the center of the galaxy. Again, under the laws of gravitation, the pull on the outer stars should be weaker than on the inner ones, and thus they should be moving slowly near the spiral edges. But the observational data shows that even these stars are orbiting at high velocities.

Similar to the conclusion made with galaxy clusters, this also clearly demonstrates that there has to be a lot more mass in the galaxy than we can see. This anomaly has been observed in every galaxy astronomers have identified. This undetectable mass that fills the galaxies or clusters shapes the motion of individual stars and galaxies through its invisible hands.

Another independent line of evidence for the existence of dark matter is found through gravitational lensing. Much like a lens bends light rays, galaxy clusters bend the light from background galaxies. The distorted galaxy images show that there is much more mass in the clusters than we detect from radiation. More mass in the cluster can make the bending very intense. The estimate of mass from this observed bending is in strong agreement with other types of observations on estimates for the amount of dark matter.

In the absence of dark matter, this ordinary matter would fly apart, because the gravity of the normal matter cannot withstand the never ending assault of dark energy. It can be said that the search for the dark matter is a "ghost hunt," as it is not detectable directly. Researchers use gravitational lensing as a tool to understand dark matter. This is a technique that utilizes the fact that the light bends in the presence of a strong gravitational field. The Hubble Space Telescope has observed strong gravitational lensing in different galaxy clusters, which underlines the presence of dark matter as predicted theoretically by astrophysicist Fritz Zwicky in 1933. While studying the dwarf galaxies, the HST provided strong evidence for the presence of dark matter. Hubble's sharp view penetrated the cosmos to see that large numbers of small galaxies remain intact, even as the bigger galaxies around them are being ripped apart by the gravitational force of other galaxies in the cluster. The halo of dark matter protects them like an invisible shield from the assault of a gravitational tug of war going on inside the clusters for several billions of years.

"We were surprised to find so many dwarf galaxies in the core of this cluster that were so smooth and round and had no evidence at all of any kind of disturbance. They must be very, very dark-matter-dominated galaxies," says astronomer Christopher Conselice, currently the principal investigator of the HST survey.

What if this "missing matter" is contributed by hitherto unknown objects? There is still a small possibility that dark matter may be baryonic in nature. The objects known as MACHOs – Massive Astrophysical Compact Halo Objects – are considered candidates for dark matter. The low-mass stars, such as brown dwarfs, are ideal contenders for such objects. However, no such brown dwarfs have been found so far.

Conversely, there is a fairly strong consensus among scientists about what is up there in the sky. Black holes and neutron stars, once thought to harbor the missing mass of the universe, are no longer the sources of speculation. If black holes or neutron stars have to be considered as candidates for the missing mass in the universe, then there must be plenty of them to account for such a large mass. The calculations show that about 90% of all the stars have to undergo supernova explosion to become black holes or neutron stars, but we know that such events are rare in the universe.

Before we continue our discussion on other dark matter candidates, let's learn about supernovae, which represent the end life of massive stars. As mentioned earlier, for astronomers supernovae are important tools in understanding the expansion of the universe.

SUPERNOVAE: THE CREATORS AND DESTROYERS OF THE UNIVERSE

Supernovae are the most energetic explosions in the universe. The power of these explosions could easily surpass the energy released by billions of nuclear warheads.

There are two basic types of supernovae – Type II and Type Ia. Observations have revealed that Type II supernovas mostly occur in spiral galaxies. The arms of these galaxies contain lots of bright, young stars. The elliptical galaxies, which are ruled by old, low-mass stars, are not the ideal place for these kinds of supernova explosions. Also, bright young stars are typically much more massive than the Sun, about 10 times the mass of the sun. This provides evidence that the Type II supernovae are produced by massive stars.

In many ways, though it sounds strange, the lives of stars are similar to that of our own life. Like us, they are born, they live, and they die. A star "lives" by fusing lighter element such as hydrogen into heavier ones in its core, where the conditions are ideal for such a process. The process known as nuclear fusion requires extremely high temperatures in order to fuse hydrogen nuclei to helium nuclei. Once the fusion process begins a star is born.

Using Einstein's famous energy-mass equation, $E=mc^2$, we can precisely calculate how much energy is released during a fusion reaction. To do this we consider four

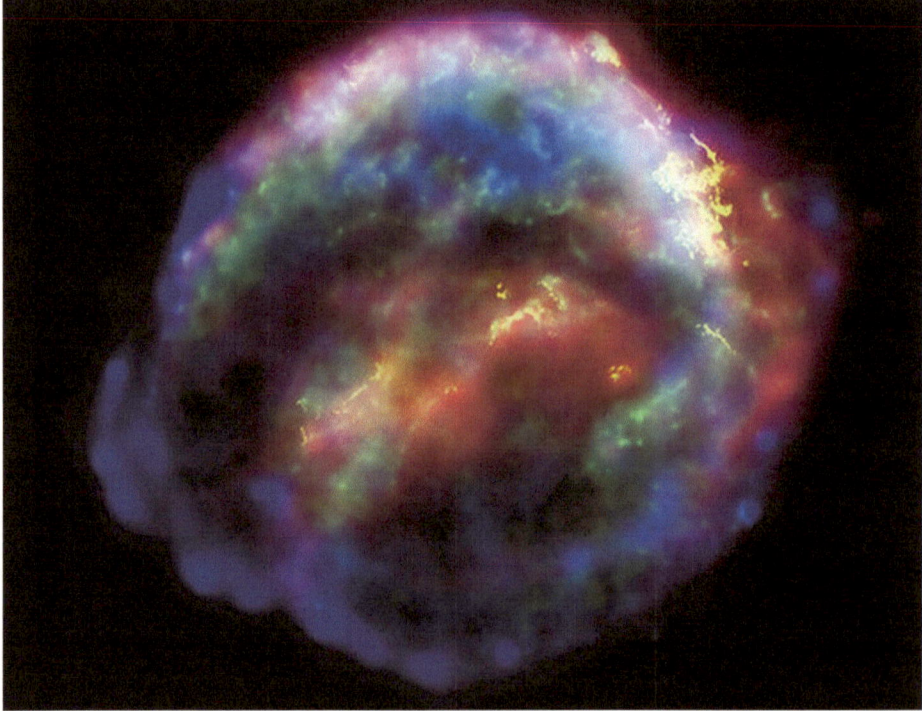

Figure 6.5. Supernovae are the grant recycling machines of the universe. This image shows expanding remains of Kepler's supernova (Photo credit: NASA).

hydrogen atoms fuse together to form a helium atom. The four hydrogen atoms together have slightly more mass than one helium atom:

$$4H \, nuclei \, weigh \, 6.693 \times 10^{-27} \, kg$$
$$1He \, nucleus \, weighs \, 6.645 \times 10^{-27} \, kg$$

The difference between these masses is the "missing mass," which is 0.048×10^{-27} kg in this case. If we plug in the missing mass in the mass-energy relation and multiply it with the square of the speed of light (c),[1] we get the energy released in a single fusion reaction. In this example, this would be equal to 4.3×10^{-12} J[2] of energy.

As you can see, the energy calculated from the above case is extremely small. Compare it with a 60 W bulb that delivers 60 J of energy every second. However, given the large amount of power output from Sun, we can assume that a large number of fusion reactions occur in the Sun in every second. In fact, it has been estimated that about 10^{38} fusions happen every second in the Sun.

The life of a star is a constant battle between two forces – gravity and nuclear fusion. While gravity tries to crush the star the pressure generated by nuclear fusion counterbalances the assault of gravity. As long as equilibrium is maintained, the star remains in good health. Eventually, nuclear fusion burns up all the lighter elements, and heavier elements

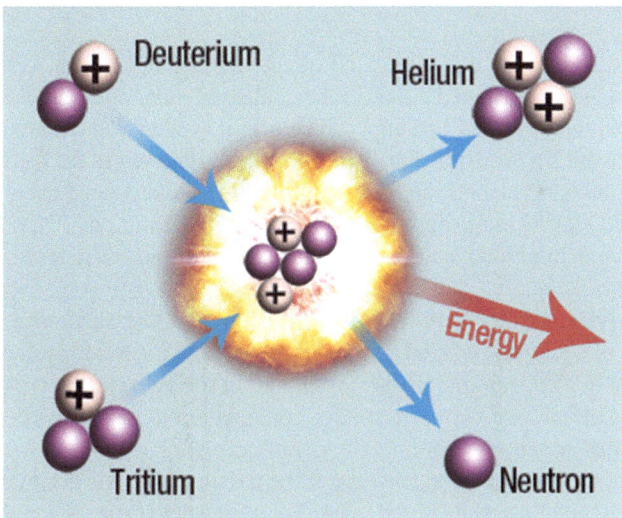

Figure 6.6. Nuclear fusion showing the isotopes of hydrogen (deuterium and tritium) combining to form helium. Energy is released in the process along with a neutron (Image credit: Wikimedia commons).

---

[1] $C = 3 \times 10^8$ m/s

[2] Joule is the unit used to measure energy or work in International System of units.

build up at the core of the star. The core may have to be at a higher temperature to fuse the heavier elements. When this process comes to a stop, the core collapses in a very brief period of time. Depending on the mass of the original star, a black hole or neutron star is formed. All but the core of the star is blown away at a tremendous speed. A shock wave ensues, and this causes the lighter elements to fuse to become heavier elements. This visual display is equivalent to several billion Suns.

Now, let's explore the end of lower mass stars. Stars such as our Sun are destined to become white dwarfs – stars that have exhausted most or all of their nuclear fuel and have collapsed to a very small size. This is a very stable state for the star, and like cosmic pebbles they quietly simmer for the rest of their lives.

Remember that most of the stars in the universe are binary systems, though our Sun is a cosmic loner. If the mass of the white dwarf star is above 1.4 solar masses a different kind of supernova explosion occurs. In the case of binary systems, the white dwarf might be able to accrete matter from a companion star or merge with another white dwarf. This pushes the mass of the white dwarf over the Chandrasekhar limit (to be discussed in Chapter 8) of 1.4 solar masses; the temperature in the core of the white dwarf will then rise, triggering explosive nuclear fusion reactions that release an enormous amount of energy. The star explodes in a few seconds, leaving no remnant. This kind of explosion is referred to as a Type Ia supernova.

"We are made of star stuff," Carl Sagan once remarked. We can see that supernovae play an important role in the creation-destruction cycles that take place in the universe. It is appropriate to say supernovae are, in fact, the most fundamental recycling events in the universe.

It is interesting to note that almost all of the elements in the universe heavier than hydrogen and helium are created either in the centers of stars during their lifetimes or in the supernova explosions described above. The newly synthesized materials are sent into interstellar space during the explosion. This new material serves as the feeding ground for next generation of stars, planets, and even life, and the great cosmic cycle continues. As Carl Sagan noted (1985) in his book *Cosmos,* "The nitrogen in our DNA, the calcium in our teeth, the iron in our blood, and the carbon in our apple pies were made in the interiors of collapsing stars."

Our Solar System might have been formed from a similar supernova explosion that marked the end of a star that existed earlier. The planets in the Solar System formed from leftover material in a disk around the proto-Sun; all of the heavy elements found on Earth (including those in our own body) must have come from the same source. This undoubtedly proclaims that we are stardust!

After the supernova explosion, the core of the parent star would become a neutron star or a black hole, depending on its initial mass. For quite some time, astronomers have proposed many arguments to account for the missing mass in the universe, including the mass in black holes and neutron stars.

WICKED WIMPS

Astronomers rule out black holes or neutron stars as the candidates for the unknown stuff that makes the dark matter, due to the fact that the supernovae that produce these phenomena are very rare. It has been estimated that a supernova explosion happens in our galaxy every 50 years or so.

However, observations from satellites and telescopes consistently underscore the existence of dark matter. The final resolution must come from the experiments. Scientific detectors are waiting tirelessly for the moment of truth.

There is no evidence of dark matter in our Solar System, because with the known mass of the planets and the Sun, we can accurately predict their motions. However, dark matter particles could be passing through our planet as WIMPS, which are being sought in experiments such as CDMS in Minnesota.

There are other exotic ideas that have been floating around to describe dark matter, such as massive neutrinos and cosmic strings. However, there is little or no evidence to back up such theories.

The announcement from the Soudan mine (2011) was the result of data collected by the detectors over the course of a year. Two pulses of possible WIMP interactions were detected. Yet, it remains unconvincing. There is a chance that the pulses were caused by the radioactivity around the detector, though extreme care had been taken to account for such effects. The experiment is buried underground to eliminate the effects of cosmic rays on detectors. Yet, even an extremely small amount of radioactivity can trigger a false signal. More precisely, there is a one in four chance of other events mimicking WIMPS in the latest findings. In coming months, more sensitive detectors might generate definitive pulses that carry the signature of WIMPS. When that happens, scientists will have a new form of matter to confront.

The next step for CDMS is a large detector known as the Super CDMS. Many scientists are hopeful of capturing definite pulses that carry the signature of WIMPS. If that happens, it would vindicate one of the wildest imaginations of modern physics – a tribute to Fritz Zwicky, who postulated the existence of dark matter for the first time in 1933 by studying galaxy clusters (1933).

What is the significance of dark matter? For one, the presence of missing matter definitely alters the data and thus our understanding of the universe. The universe was formless and void in the beginning. What prompted the galaxies to form in an evenly distributed early universe?

It is assumed that dark matter provided the initial gravity for the clumps to take shape. In the absence of such initial clumping, galaxies or planets would not have been here. Thus dark matter played a crucial role in the formation of such systems and, for that matter, our own creation. So the glimpse of WIMPS may lead us back to our cosmic ancestry.

While men on the ground are preparing the detectors for a glimpse of WIMPS, manmade machines in space have located the halo of the dark matter enveloping the galaxies. The Hubble Telescope and Chandra X-ray Observatory have played critical roles in this regard. Using the data, researchers were able to map dark matter within the galaxies and in clusters. These telescopes have captured the titanic galaxy-cluster collisions. These observations indicate that ordinary matter separated from dark matter as a result of such collisions.

Some physicists stretch their imaginations even further. They propose creating the matter in a lab instead of passively observing it. Assuming that much of the dark matter was created in the very first moments of the Big Bang, the machines that mimic such scenarios might be able to produce dark matter. The Large Hadron Collider at CERN is one such machine capable of creating similar conditions that existed in the Big Bang era. So far, no hints of dark matter have been observed.

Again, some physicists call for modification of Newtonian laws of physics in order to deal with dark matter. In physics, Modified Newtonian Dynamics or MOND is a theory proposed by physicist Moti Milgrom as a solution to the missing mass problem. It advocates modification of Newton's law of gravity to explain the galaxy rotation problem that demanded the need for more mass. Milgrom demonstrated that if a modified version of Newtonian dynamics is used to describe the motions of bodies in a gravitational field (of a galaxy, say), the observational results are reproduced with no need to assume hidden mass in appreciable quantities (1983).

The assumption is that at large scales, the laws of gravity are different from the known laws of gravity that we use lavishly to study the universe. The Newtonian laws predict that the orbital velocities of planets must decrease as a function of distance from Sun. However, that was not the case with rotational galaxies when their velocities were observed. MOND implies the need to change the law to describe the cosmological evolution, in terms of the material content of the universe, in the limit of smaller accelerations.

However, in recent years, an ever-increasing collection of measurements have been performed and produced data that support the existence of dark matter in our universe (Bertone et al. 2005). Perhaps, dark matter presents one of the most compelling direct indications of new physics beyond the standard model that describes the interactions of particles in normal matter. Along with dark energy, dark matter provides a window of opportunity to explore the unknown as well as to study the exciting new physics. A complete understanding of dark matter requires utilizing almost all branches of physics and astronomy. Although we have yet to detect this strange material in the laboratory, there is a great deal of evidence that points to the necessity of its existence.

As the experiments continue around the world for the detection for some of the most tantalizing stuff in science, it also draws our attention to how our minds and imaginations are gifted for knowing the unknown. We need dark matter for the structures of our universe to exist, yet we don't know much about it. The detection of dark matter will prove that we are completely different from the universe we inhabit, a contradicting proposition to our current understanding. If we don't find dark matter, however, it will offer us an even deeper mystery about the unknown forces of gravity we have already detected.

As Vera Rubin, of the Carnegie Institution of Washington, whose pioneering work established the presence of dark matter in galaxies, once said in 1966: "In a spiral galaxy, the ratio of dark-to-light matter is about a factor of ten. That's probably a good number for the ratio of our ignorance-to-knowledge. We're out of kindergarten, but only in about third grade."

As evidence for the existence of dark matter and dark energy mounts up, we may have to accept the fact. Until then, we must be prepared to live in ambiguity, and that is quite customary in science. It is better to live with an unknown rather than embracing a false reality.

Figure 6.7. Dark matter halo. This artist's impression shows the Milky Way Galaxy surrounded by dark matter. The blue halo of material surrounding the galaxy indicates the expected distribution of the mysterious dark matter, which was first introduced by astronomers to explain the rotational properties of the galaxy and is now also an essential ingredient in current theories of the formation and evolution of galaxies (Image credit: European Southern Observatory).

REFERENCES

Adam, G. R., et al. (1998). Observational evidence from supernovae for an accelerating universe and a cosmological constant. *The Astronomical Journal, 116*(3), 1009–1038.

Ahmed, Z., et al. (2011). Search for inelastic dark matter with the CDMS II experiment. *Physical Review D, 83.* arXiv:1012.5078.

Alwall, J., & Tandean, J. (2013). Heavy chiral fermions and dark matter. *Advances in High Energy Physics, 2013.* Article ID 915897, 15 pages. doi:10.1155/2013/915897.

Armendariz-Picon, C., Mukhanov, V., & Steinhardt, P. J. (2000). Dynamical solution to the problem of a small cosmological constant and late-time cosmic acceleration. *Physical Review Letters, 85,* 4438–4441.

Barkana, R., & Loeb, A. (1999). The photoevaporation of dwarf galaxies during reionization. *Astrophysical Journal, 523*(1), 54–65.

Bertone, G., Hooper, D., & Silk, J. (2005). Particle dark matter: Evidence, candidates and constraints. *Physics Reports, 405*(5–6), 279–390.

Chandra:: Field guide to x-ray sources:: supernovas & supernova remnants. Available at: http://chandra.harvard.edu/xray_sources/supernovas.html. Accessed 2 Jan 2012.

Garrett, K., & Dūda, G. (2011). Dark matter: A primer. *Advances in Astronomy, 2011.* Article ID 968283, 22 pages. doi:10.1155/2011/968283.

Hubble, E. P. (1954). *The nature of science, and other lectures* (p. 6). California: Huntington Library.

Milgrom, M. (1983). A modification of the Newtonian dynamics as a possible alternative to the hidden mass hypothesis. *Astrophysical Journal, 270,* 365–370.

Mitchell, S. (2006). *Bhagavad Gita: A new translation*. New York: Three Rivers Press.

Olga, M., et al. (2005). *Constraining inverse curvature gravity with supernovae*. Available: http://lss.fnal.gov/archive/2005/pub/fermilab-pub-05-466-t.pdf. Last Accessed 04 May 11.

Relativity and gravitation . *41*, 207–224. Available at: http://link.springer.com/article/10.1007%2Fs10714-008-0707-4. Accessed 05 June 2010.

Riess, A. G., Filippenko, A. V., & Challis, P. (1998). Observational evidence from supernovae for an accelerating universe and a cosmological constant. *The Astronomical Journal, 116*(3), 1009–1038.

Rubin, V. C. (1983). Dark matter in spiral galaxies. *Scientific American, 248*(6), 96–108.

Rubin, V. (1996). *Bright galaxies, dark matters* (p. 147). New York: Springer Verlag.

Sagan, C. (1985). *Cosmos*. New York: Ballantine Books.

Science Magazine (2012). Available at: http://www.sciencemag.org/content/327/5973/1619.full. Accessed 27 Mar 2012.

Steinhardt, P. J. (1997). Cosmological challenges for the 21st century. In V. L. Fitch & D. R. Marlow (Eds.), *Critical problems in physics*. Princeton/New Jersey: Princeton University Press.

Steinhardt, P. J. (2003). A quintessential introduction to dark energy. *The Royal Society, 361*(1812), 2497–2513.

Steinhardt, P. J., & Turok, N. (2002). A cyclic model of the universe. *Science, 296*, 1436–1439.

Zlatev, I., Wang, L., & Steinhardt, P. J. (1998). Quintessence, cosmic coincidence and the cosmological constant. *Physical Review Letters, 82*, 895.

Zwicky, F. (1933). Spectral displacement of extra galactic nebulae. *Helvetica Physica Acta, 6*, 110–127.

Zwicky, F. (1937). On the masses of nebulae and clusters of nebulae. *The Astrophysical Journal, 86*, 217–246.

7

Is the Universe Infinite?

FINITE OR INFINITE?

> *Innumerable Suns exist; innumerable earths revolve around these Suns in a manner similar to the way the seven planets revolve around our Sun. Living beings inhabit these worlds.*
>
> – Giordano Bruno, a controversial figure of the Italian Renaissance who was burned at the stake for heresy and pantheism in 1600.

The concept of infinity, all of us know, is not something that is easily understood. Let us begin with a 1975 short story by the famous Argentine writer Jorge Luis Borges (2007).

In "The Book of Sand," Borges narrates the story in which he meets a Bible seller. He wants to sell Borges a clothbound book that he bought in India called *The Book of Sand.* "Neither the book nor sand has beginning or an end" says the seller to Borges. Upon examining the book, Borges finds that the pages in the mysterious book are infinite. He can't find the first or last page. The book is written in an unknown language occasionally punctuated by illustrations, and the text is cramped and arranged in vesicles. As he turns the pages more pages seem to spring from the book. He buys the book and places it on a bookshelf.

Over time, Borges' obsession with the book grows alarmingly. He refuses to go outside for fear of theft. His obsession with the book is making him a prisoner. Soon he realizes that the book is a monster – a nightmarish object that is destroying his own life. He thinks of burning it, but fears the possibility that the smoke from an infinite book will suffocate the whole world. So Borges decides to hide it among the basement bookshelves of the national library. He reasons that "the best place to hide a leaf is in a forest."

Like in Borges story, are we troubled by the immensity of the universe? The history and nature of the universe as deciphered by humans, based on the information we have, is

S. Mathew, *Essays on the Frontiers of Modern Astrophysics and Cosmology*, Springer Praxis Books, DOI 10.1007/978-3-319-01887-4_7, © Springer International Publishing Switzerland 2014

only a small part of the big picture. It is a universe of our own creation and one that is plagued by dark forces and our own ignorance.

It is quite normal to wonder about the universe. Its vastness and majesty trigger awe. But its nature is still a lingering mystery. Is the universe finite or infinite?

When we talk about the universe, we mean everything. In fact, there is much beyond what we, or even our most advanced telescopes, can see. Researchers can trace the history of the universe to any point since the first light penetrated the fabric of the cosmos and reached us. However, there is a fundamental limit to this conundrum. We do not and will not have any idea about any of the regions in the universe from where the light, carrying valuable information that helps us to understand the history, came.

Our most acceptable model of the universe says it has a finite origin but is expanding. Though it has enormous volume that we can perceive, it remains finite now, and that volume will increase, so only in the infinite future will it actually be infinite. It opens up the question about the geometric shape of our universe. Is it flat like a sheet of paper, closed like a football, or an open entity without any limits? Actually, there are conflicting answers to this question even as cosmologists scan through the huge data they have been collecting to nail down this perplexing issue.

In an infinite universe, a photon (light) will never find its way back unless it is forced to do so. But in a finite universe, it must end up in the same location from where it started. The density of matter in the universe would play a critical role in deciding which of these is correct. Even now, astronomers are not completely confident about either of these possibilities.

Yet, it is completely within the limits of scientific inquiry to think about the unknown. So what lies beyond the known borders of the known universe? Even if light began its voyage immediately after the Big Bang, it could have so far traveled only about 13.7 billion light-years of distance. To make matters worse, the same light has to encounter an expanding universe during its trip, making it unable to reach us. Are there any structures and patterns out there that are like or unlike what we have in our observable universe?

Though we can't ever see such configurations, there is a way that we could feel the effect of such entities if they do exist. That's exactly what astrophysicist Alexander Kashlinsky of NASA's Goddard Space Flight Center suggests (2008). By studying the giant clusters of galaxies and their scattering, he and his team noticed a peculiar phenomenon. They concluded that something from way beyond the edge seems to be pulling powerfully on galaxies in our universe, yanking them along in a motion he calls "dark flow." This has nothing to do with the known effects of dark energy that causes the galaxies to move away from each other. These researchers say there's evidence that galaxy clusters are being pulled along by a force outside the visible universe.

As if dark matter and dark energy are not enough to complicate the mystery surrounding the universe, the new phenomenon of dark flow, though many scientists disagree, has added another dimension to the already incomprehensible picture of the cosmos.

In other words, our galaxies are pulled by matter that is beyond our known universe. Some researchers even suggest that this could be from other universes that float like bubbles in the cosmic ocean along with our own, unknown to us. At this point, such thoughts do not merit any serious consideration, as they lack independent observational evidence.

Nonetheless, searching for the edges our cosmos remains a satisfying endeavor. Such thoughts were expressed long ago as illustrated in the words of Marcus Aurelius, Roman emperor and stoic philosopher: "He who does not know what the world is does not know where he is, and he who does not know for what purpose the world exists, does not know who he is, nor what the world is."

Is there anything special about our location in the universe? According to the Copernican principle, which is the foundation of modern astronomy, it must not be so. Whatever we experience in this part of space-time must be more or less the same everywhere even if we can't observe it. This must be true for any observer anywhere in the universe. The inability to discern the parts of the universe gives rise to the impression that the universe must be infinite in every direction we look.

The universe may be infinite, but its origin, as we understand it, demands that it has a finite age. All we can say is there is much more than what we see out there.

The concept of infinity is vividly described in Vedic cosmology. In Sanskrit, *ananta* means infinity, and the universe, though cyclic in nature, is anaadi[1] and anant, which implies something without beginning and end. Hinduism subscribes to the concept that the universe had no beginning and has no end. It also maintains that there are many more worlds and universes than there are drops of water in the ocean. These universes were made by Lord Brahma, the Creator, maintained by Lord Vishnu, the Preserver, and destroyed by Lord Shiva.

After the destruction of each universe, there will be a vast ocean left. Lord Vishnu, resting on the great snake Ananta,[2] floats on this ocean. Some scriptures suggest that a lotus flower springs from his navel, and from this emerges Lord Brahma. And it is from Lord Brahma that all is manifest, to be destroyed eventually.

Any discussion about the nature of the universe underscores human limitations and ignorance. As Einstein once remarked, "Only two things are infinite, the universe and human stupidity, and I'm not sure about the former."

As far out as our modern telescopes can see, billions of stars group together to form galaxies, galaxies bundle together to form clusters, and clusters clump into super clusters. They all seem to hide a message coded in their unique pattern, called fractals. Some believe this is the message from the cosmic sculptor for sentient beings like us, while others think it is the ultimate truth expressed in geometric language for us to explore. Some see beauty in it while others seek truth from it. So let's discuss about fractals before we proceed.

A WORLD OF WORLDS

Fractal patterns have no beginning or end. The whole exists in parts, and a part exists in the whole. The universe is in you, and you are in the universe.

When cosmologists attempted to map the configuration of the universe by recreating past events and extracting information from radiation, they found that the galaxy distribution in the universe followed a certain fractal pattern. Fractals are like a whole world on a smaller scale embedded inside a world of increasing complexity. They are the smaller versions of the relatively bigger and similar structures – a reduced copy of the original.

In the pinnacle or creation, it seems that the forces of nature left out an important clue for us to hunt down the secrets of the universe. Are fractals the road map to reality?

---

[1] Without a beginning.

[2] Also refers to the serpent on which Vishnu lies.

Do they symbolize a deliberate attempt by nature, or do they just exist in the eyes of the beholder?

Surprisingly, fractals can be seen not just in celestial objects but in terrestrial objects, too. For example, look at a sunflower or broccoli. You will see that a smaller copy of the whole is embedded successively and, if you keep looking, you can find smaller and smaller copies. The object in larger scale is composed not just of different parts, but by tiny whole ones. The unique nature of fractals is that smaller versions of an object are continuously repeated to generate the bigger object in a manner so that the small and big preserve all the features. They are small in big, big in small, part in whole, and whole in part. The enlightened might say there is, in fact, no such distinction as alpha and omega – the universe is in you, and you are in the universe.

Fractals could be described as the finite holding the infinite. From snowflakes to shorelines, from clouds to seashells, from blood vessels to ocean waves, all these seemingly irregular shapes harbor a beauty beneath their intricate systems. The fractal patterns are everywhere – in human lungs and kidneys, and in the rhythm of the heart – in the essence of life. Could they be the heartbeat of the cosmos as well?

The Sloan Digital Sky Survey (SDSS) is one of the most ambitious and significant projects in the history of astronomy. The goal of this mission was to create a true picture of the universe by mapping millions of galaxies and quasars that govern our cosmos. When the data was released on the distribution of thousands of galaxies they studied, the results were clear and compelling. Most physicists agreed that on a relatively small scale the galaxy distribution follows a fractal pattern. Like in the humble sunflowers, one can witness how the fractals unfold in the majesty of the swirling galaxies.

Figure 7.1. The fractal universe. Is matter in the universe distributed in a fractal pattern? (Image source: Creative Commons)

The infinitely repeated similarity is the key to fractal generation. In nature, this is expressed in everything from meek objects to mighty galaxies. Some consider it a universal truth. It is not just the aesthetic feature of the fractals that makes them significant, but they are the key to understanding order in disorder or beauty in anarchy, wherein chaos disappears to create serenity.

Scientists and mathematicians often bury the beauty of nature in abstract mathematical symbols, anything from alpha to omega. Traditionally, they love to describe smooth and definite patterns with their equations and shy away from complex and irregular patterns. In that sense, Benoit Mandelbrot, who died in 2010, was a radical mathematician. He went on to explore the hidden beauty beneath seemingly chaotic systems.

He proposed that the complexity we observe in nature can be generated by iterations of a simple entity over a large number of repetitions. In other words, the fundamental equation that represents the smallest link could repeat endlessly to create a complex system. The iterations can have fractal dimensions instead of the usual whole-number dimensions. Fractal geometry ideas are used in a variety of fields from geology to cosmology and from medicine to the stock market.

Mandelbrot went on to explore the underlying geometry and mathematics in varieties of systems that have been largely ignored by conventional mathematicians. His work was initially confined to smaller and familiar terrestrial objects, but it has been extended to systems that create chaos, such as weather or stock markets. Though the prediction of the behavior of such systems is still in its infancy, it is not as farfetched as once thought. Fractals reveal that unpredictability and complexity are part of nature, and that they are not undesirable.

The Fractal Geometry of Nature is a mathematical text published in 1983 by Mandelbrot that addresses many of the mathematical puzzles that involve fractals. He began originally as a researcher at IBM and later became a faculty member at Yale University. Fractal geometry ideas are used in a variety of fields, from geology to cosmology and from medicine to the stock market. It explores underlying geometry and mathematics in varieties of systems that have been ignored largely by conventional mathematicians.

Though exact predictions of the behavior of such systems are still in their infancy, they are not doomed as once thought. Fractals reveal the fact that unpredictability and complexity are part of nature, but they are not an undesirable part.

The ability of fractals to derive order from disorder is at the heart of chaos theory, a new and exciting science. Though chaos implies disorder and unpredictability, such as the weather system, scientists glimpse a hope here as chaos could be, after all, deterministic.

To demonstrate the relation between the outcome and its severe dependence on initial conditions, it has been said that a butterfly flapping its wings in South America can affect the weather in Central Park. But, it has also been said that even accidents follow fixed laws. Are we yet to discover these laws?

The slow and continuous evolution of intelligence and complexity from single cells could also be rooted in a similar process that began, like cosmic creativity, long ago. The ability of fractals to create intricacy from relatively simple states we are probably evolving our own creativity and intelligence. Perhaps, those fractals that are veiled underneath the clouds, mountains, and coastlines and also in the mighty cosmos, are carrying out their objectives to reach the pinnacle of creation as we read this.

Engineers can model wonderful structural designs with common geometric shapes, such as lines and circles, and with software they can generate 3D models of the same. A mathematician can create a set of equations that would describe this model. Ask them to model a cauliflower or broccoli? They would probably have stumbled, at least until the work of Mandelbrot became an established branch of geometry.

What is the best approach to decipher the secrets of the universe? Mathematical equations, as the classical physicists thought, or pictures and patterns? Mandelbrot favored the second choice. While physicists such as Isaac Newton thought God was a mathematician and believed that the secrets of nature are encoded in mathematical equations, fractals teach us that the blueprint of reality lies in shapes and molds. Also, that chaos and complexity are to be celebrated rather than ridiculed, because they are the pure manifestation of an underlying simple truth.

The near-uniform radiation left over from the birth of the universe suggests that it began in a void and featureless state. How did the formation of galaxies or distribution of matter in the universe follow a unique pattern? Without a proper theory to explain it, researchers are like acrobats without a net. As the Sloan Digital Sky Survey continues its efforts to cover more galaxies, many express optimism that a clear picture will eventually emerge. This could lead us to a higher level of comprehension about the cosmos and, even more importantly, about ourselves.

At present, it is not clear whether fractal patterns can be observed in galaxy distributions of any distance in the universe. But a small team of physicists argue that recent data show the universe continues to look fractal as far out as our telescopes can reach.

If large-scale fractals do exist throughout, what might be the implications? Like the galaxies that exhibit fractals, our universe could be pictured like a Russian nesting doll, which metaphorically denotes the recognizable relationship of "object within similar object." Our universe could be residing in a bigger version of its own, and many smaller versions of it could be embedded within our own universe. Does this fractal nature of the universe make sense?

Some cosmologists believe that even the near-featureless state of our cosmos in the beginning – the one currently so considered – was preceded by something simpler still, perhaps little more than rapidly expanding empty space. In a process spanning billions of years, this fundamental similarity multiplied numerous times to create our present state of the complex universe. If the fractal hierarchies are ubiquitous, as some like to think, we have no choice but to be confined within a series of universes with increasing magnitude and intricacy – a different version of the multiverse that already has its own niche in modern cosmology.

The infinite hierarchy of worlds within worlds may not soon become the dominant theory in astronomy. Yet, it is not unfamiliar to Hindu philosophers. During the discourse in Kurukshetra[3] with Krishna,[4] Arjuna[5] saw the entire universe, divided in many ways, but

---

[3] According to Mahabharata, the epic war between good and evil happened at Kurukshetra. Also, the Bhagavad Gita was preached here during the war.

[4] An incarnation of Vishnu and he acted as Arjuna's charioteer during the Kurukshetra war.

[5] One of the greatest warriors and a hero in the Mahabharata. The discourse between Krishna and Arjuna is the content of the Bhagavad Gita.

standing as (all in) One (and One in all) in the body of the God of Gods. The Upanishads proclaim that the self is one, though it appears to be many. This self is the truth, and this true self is exactly the same as Brahma, which is omnipresent and is manifested in nature (prakriti) and unmanifested as consciousness. This self is not the body, nor is it the mind, nor is it the ego. With this self that is the same in all things, there can be no sense of "I" or "mine," as all are one and the same.

We will, no doubt, ultimately discover the law of fractals, because its dance is the same everywhere – in you and in the cosmos. If fractals represent the patterns in the universe, the force that makes it possible is gravity, the all-pervading force of the universe.

The patterns exist not only in visible nature. Mathematicians figure out the patterns in numbers. Following is a remarkable story of a mathematician who overcame hardships in his search for mathematical patterns.

THE MAN WHO DEFINED THE UNDEFINED

"The mathematician's patterns," G. H. Hardy wrote, "like the painter's or the poet's, must be beautiful; the ideas, like the colors or the words, must fit together in a harmonious way. Beauty is the first test: there is no permanent place in the world for ugly mathematics."

However, mathematics has the peculiar reputation of being a dull and dry subject that is enjoyed, if at all, by a few. While the symphonies of Beethoven and Mozart are appealing to all, the mathematical theorems are not so lucky; they are dismal and cruel to most of us. No surprise, Gauss and Euler are not household names.

In the list of the greatest mathematicians ever, there is a self-taught genius who lived only 32 years, 4 months and 4 days on this planet. Yet, his findings are so profound and elegant that many narrate them as the creative efforts to reveal the symphony of numbers. Sreenivasa Ramanujan was born to a poor family on December 22, 1887, in Erode, a city in Madras State (now Tamil Nadu) in South India and passed away on April 26, 1920.

Ramanujan's ardent passion for numbers began early in his school years. His first works include creating magic squares of different sizes. His notebook gives the examples of magic squares of different dimensions, up to 8×8, and suggests some general method to construct them. A magic square is a square array of distinct natural numbers so that the sum of the numbers in each row, column, or diagonal is the same. Nowadays, with the help of *Mathematica* or similar tools, a system of equations can be solved easily to find the numbers to fit in a magic square.

6	1	8
7	5	3
2	9	4

Figure 7.2. The man who defined the undefined. Ramanujan sought to reveal the beautiful patterns that exist in number systems (Source: CC Wikimedia).

Above is a 3×3 magic square from Ramanujan's notebook. The elements in the middle row, middle column, and each diagonal are in arithmetic progression and all columns, rows and diagonals add up to 15.

This, of course, is recreational mathematics, and anyone can create similar magic squares by choosing different numbers in arithmetic progression. In addition to such elementary mathematical problems, the notebooks contain about 3,000 interesting and profound theorems that deserve the attention of mathematicians. These theorems are beyond elementary mathematics such as the magic square or the partition of numbers.

Most of the discoveries recorded in these notebooks are without the proofs necessary to reach the final result, a frequent criticism against the Ramanujan approach.

"The discovery of his 'Lost Notebook' caused roughly as much stir in the mathematical world as the discovery of Beethoven's tenth symphony would cause in the musical world," remarked Bruce Berndt, who along with G. E. Andrews published the first of several volumes of books in which they provided proofs for the formulas found within the famous notebook.

As a result of his unusual devotion to mathematics, Ramanujan failed most of his other classes when he was in college. As a poor college dropout, Ramanujan had to live with the support of his friends, but he continued to work on mathematical problems.

He used a slate and chalk to do the calculations and only recorded the final results in notebooks. He allegedly told a friend, "When food itself is a problem, how can I find money for paper? I may require four reams of paper every month."

Ramanujan passed along many of his interesting theorems to mathematicians in English. His uncanny insights drew the attention of the renowned English mathematician G. H. Hardy, who helped him to go to Cambridge University.

Ramanujan set off for England in 1914, and the next 5 years he worked with Hardy on many exciting mathematical projects. This well-celebrated collaboration between Ramanujan and Hardy is one of the most colorful chapters in the history of mathematics. Robert Kanigel's book *The Man Who Knew Infinity: A Life of the Genius Ramanujan,* records the astonishing story of Ramanujan and his collaboration with Hardy.

Unfortunately, Ramanujan's health deteriorated, and he became seriously ill. Hardy visited him in the hospital. "I rode here today in taxicab number 1729," Hardy told Ramanujan. "This seems to be a dull number, and I hope it's not an unfavorable omen." To which Ramanujan replied, "No, it is a very interesting number; it is the smallest number expressible as the sum of two cubes in two different ways."

$$1,729 = 1^3 + 12^3 = 9^3 + 10^3$$

In mathematics, this is the smallest number in the family of so-called taxicab numbers that derived their name from the above incident.

Ramanujan returned to India in 1919, but he died the next year. He defied poverty and transcended national borders and cultures to define the undefined. If patterns of numbers can create a symphony, Ramanujan is the Beethoven of number theory.

We agree on the fact that nature prefers patterns. What holds a pattern together to make a physical structure? We have to go back to gravity to know that.

GRAVITY: A COSMIC GLUE?

In one of the space shuttle missions shuttle (Atlantis' STS-132 mission 2010), as a symbolic tribute to the father of gravity, NASA astronaut Piers Sellers carried along with him a chip of wood from Isaac Newton's legendary apple tree that helped Newton formulate the law of gravity.

Whatever the real story connected to this tree, one thing is certain – the force that attracts the apple to Earth is the same force that keeps the planets in their orbit, stars in the galaxies, and galaxies in the universe. Without this cosmic glue, there would be no stars, planets, or even life. Newton identified the force, and Einstein tamed it, yet this very first force of nature harbors secrets that we may never know. In a post-Einsteinian world, the attempts to untangle this force have introduced some inconceivable effects that shatter many of our deeply held convictions.

Gravity is the first force we understood quantitatively, thanks to Newton's law of gravity. Even our rocket science is based on this simple and powerful law of nature expressed mathematically by Newton.

Though Newtonian equations offered functional insight into this force, Newton wasn't sure – rather, he didn't bother – about how gravity worked. It took another 200 years and the genius of Einstein to unlock that secret. For him, gravity was not a force due to mass as Newton envisioned, but a property of the geometry of space-time. According to Einstein, massive objects bend the surrounding space-time to create the effect of gravity, which is a huge transition from the Newtonian worldview.

The seemingly familiar force of gravity that is so integrated into our psyche can appear in totally unfamiliar ways. Black holes are examples of how gravity can manifest itself in its extreme. The massive stars that collapse to become black holes defy any conventional explanations available in science. The gravity in black holes is so powerful that even light doesn't escape from this weird region of space. In fact, the indirect detection of black holes becomes possible only due to the immense gravitational effect on the surrounding stars.

Consequently, this gravitational force has an effect on time as well, as predicted by Einstein's theory. Time must slow down in such a strong gravitational field. In other words, if we could observe an object falling into a black hole, it would be frozen in time – the ultimate slow motion.

The Pioneer anomaly is another example of our failure to comprehend gravity. The Pioneer spacecraft launched in the early 1970s are now billions of miles from Earth, beyond even our Solar System. NASA discovered that an inexplicable force appears to be acting on the probes as they head away from the Sun. They seem to be moving differently from how they were predicted to move. Scientists have yet to determine the cause of this violation of our known laws of gravity.

One of the foremost challenges facing twenty-first century physics is to reconcile the theory of relativity, which explains large-scale gravity, to quantum mechanics, the world of fundamental particles.

Scientists agree that four fundamental forces govern nature – namely, gravity, electromagnetic force, and the strong and weak forces. Quantum mechanics asserts that there is a particle associated with each of these forces. As a familiar example, light, an electromagnetic force, is carried by the photon, the fundamental particle that governs electromagnetic force. So far, the carrying particle for gravitational force, namely, the graviton, has not been found, though theoretical models predict its existence.

Einstein's theory predicts the possible existence of gravitational waves similar to the ripples in a pond that travel from the point of their origin. Thus gravitational waves must originate as ripples in the fabric of space-time, as a result of cosmic events such as a star collapse or galaxy collisions. Unlike electromagnetic waves, these gravity waves are so delicate, they need the most sensitive instruments to detect them.

Gravity wave detection is the objective of LIGO (Laser Interferometer Gravitational-Wave Observatory). Two of the biggest such interferometers in the United States – one in Hanford, Washington, and the other in Livingston, Louisiana – are collectively called LIGO. These systems are sensitive enough to detect the ripples in the space-time fabric caused by cosmic events, such as supernova explosions, and transmitted by gravity waves. These waves carry information on the events, such as the explosion and collision of stars or even the origin of our universe. It is estimated that the detection of gravity waves will be possible in coming years.

This omnipresent force, though, never stops straining our imaginations. From string theory to quantum gravity, the struggle to understand this force continues. Some theories even shake the foundation of its very existence.

Professor Erik Verlinde, a string theorist at the University of Amsterdam, in a recent paper, "On the Origin of Gravity and Newton's Laws of Motion," (2011) argues that the way science has been treating gravity is wrong. He asserts that there is something more basic, and gravity is just an emergent phenomenon of those basic constituents: "Think of the universe as a box of Scrabble™ letters. There is only one way to have the letters arranged to spell out the Gettysburg Address, but an astronomical number of ways to have them spell nonsense. Shake the box, and it will tend toward nonsense, disorder will increase, and information will be lost as the letters shuffle toward their most probable configurations. Could this be gravity?" (NYTimes.com 2010).

Science demands that the theory of gravity respect the laws of the quantum world. Matter at the fundamental level obeys the laws of quantum mechanics, where the effect of gravity is small, and gravity follows the theory of relativity, where the quantum effects are insignificant. But black holes teach us that both effects are profound, as they are small in size but huge in mass. A complete comprehension of gravity will indeed be a big leap for science.

We have all been taught that everything falls because of gravitational force. We see this force make and break the constituents of the universe. We know it gives and takes. It has a profound impact on our daily lives. We owe our existence to this force. What we experience as our weight is the result of this force of attraction on our body by the planet Earth. We are aware that our mechanical and biological clocks tick differently in different gravitational fields. We can shoot rockets by overcoming this force, yet our knowledge of gravity remains in its infancy.

We are not in a position to comprehend the size of the universe; neither are we in a position to cross the vast cosmic ocean. But we will continue our attempts as curiosity takes us on explorations through the universe, as we have done in the past. The history of the Voyager spacecraft teach us the same.

"GOOD-BYE TO THE SOLAR SYSTEM," SAYS VOYAGER

Less than 2 years ago, the Voyager 1 spacecraft had just started crossing our Solar System. The veteran probe is currently more than 10 billion miles from us and adds a million miles every day to its trip. Yet, the most distant manmade object is not bidding farewell to us, really, as it continues to send information that help us to better understand the cosmos.

Voyager 1 and Voyager 2, launched in 1977, were originally designed to conduct flyby studies of Jupiter and Saturn. But these spacecraft eventually went on to study all the four outer planets and many of their moons. They follow special trajectories at different speeds, with Voyager 1 at a higher speed, which has reached a distant point at the edge of our Solar System.

The Voyager spacecraft carry messages intended to communicate our story to any other civilization, should that be encountered. Known as the Golden Record, this message is carried by a phonograph record, a 12-in. gold-plated copper disk containing sounds and images selected to portray the diversity of life and culture on Earth.

Each record contains multilingual greetings, music, pictures, and a variety of natural sounds. One of the greetings reads, "This is a present from a small, distant world, a token of our sounds, our science, our images, our music, our thoughts, and our feelings. We are attempting to survive our time so we may live into yours. We hope someday, having solved the problems we face, to join a community of galactic civilizations. This record represents our hope and our determination, and our good will in a vast and awesome universe."

The 33-year-old message meant for other civilizations has so far had no takers. In fact, it was not meant to be a serious attempt to begin an interstellar communication. This was

Figure 7.3. Each of the two Voyager spacecraft launched in 1977 carrying a 12 in. gold-plated phonograph record with images and sounds from Earth – a message from Earthlings (Image credit: NASA/JPL-Caltech).

more of a symbolic scientific adventure as the message and the probe will take 40,000 years to pass within 1.6 light-years of another star, which is not even considered to have a solar system.

The Voyager journey teaches us how limited our physical reach in the universe is. Despite the limitations and challenges we face, we continue to explore the universe.

THE DAWN OF EXOPLANETS

Astronomers are finding numerous planets beyond our Solar System. Like dewdrops on a spider web, they dangle on the cosmic web stretching to the known borders of space and time. They come in different size and masses, but the glory of their host star conceals them from us. They are the extrasolar planets, or exoplanets – planets belonging to other solar systems.

Exoplanets have sparked the curiosity and imagination of humankind about other Earths and have been aptly labeled as other worlds. In the ongoing race to find more of them, all agree, it is only a question of time before we encounter a true twin for our home planet. Then what? Such a game-changing discovery will demand serious debate on the meaning of our existence and even life.

In the past, exoplanets were the culmination of the dreams and imagination of many astronomers. Even when scientists first understood that the universe is filled with billions of stars and their galaxies, the number of known planets in the cosmos remained just a handful – the eight in our Solar System, which have been known to us for a long time.

The planets around stars such as the Sun were mere speculation and a distant scientific possibility. This ended in 1995, when astronomers announced the confirmed discovery

(Mayor, M., and Queloz, D.) of the first planet outside our Solar System. Swiss astronomers Michel Mayor and Didier Queloz of Geneva discovered a planet orbiting the star named 51 Pegasi. This planet was at least half the mass of Jupiter and definitely not a rocky planet like ours, not in the comfort zone of a star.

However, since that discovery, in the past 15 years, the number of exoplanets has increased dramatically and currently stands above 500 and still counting. Surprisingly, most of them are gas giants and are bigger than Jupiter, the largest planet in our Solar System. None of them are Earth-like, and all are located outside the so-called habitable zone of their host star. A planet can survive only in a narrow zone. Too close to the parent star, and it will be vaporized; too far, and it will exist as a frozen world. But a planet cannot survive without a parent star. The parent star commands respect and demands that the planet live at a healthy distance to be a dynamic and evolving planet like Earth, where something unique can happen and evolve – life.

In September 2010 researchers announced the discovery (NASA 2010) of a planet somewhat similar in size to Earth in a potentially habitable zone of a star. The global media and scientific worlds were abuzz with this discovery of a planet that orbits a red dwarf star, known as Gliese 581, the most common type of star in our Milky Way Galaxy. Red dwarf stars don't have the firepower of main sequence stars like the Sun, but they can still harbor and provide life to planets like this newly found one.

Figure 7.4. Artist's impression of the newly discovered planetary system Gliese 581. The Gliese 581 planetary system is now known to have at least four planets (Image credit: ESO).

Fanciful tales about possible life forms on such a planet filled the air even though astronomers remain skeptical. Though it was confirmed that this planet, which is 20 light-years away from us, is in a Goldilocks zone, the so-called not too cold or hot zone from the star, ascribing any more qualities to it is beyond the true spirit of science. Some have dubbed it as the first habitable-zone planet, although others are discomforted by such assertions.

However, the significance of the discovery is not limited to finding a planet in a habitable zone. It is that it provides evidence for the potential number of such planets in our galaxy or, on a larger scale, in our universe. Just a decade ago, such an object had been simply theoretical in nature.

Astronomers use techniques such as Doppler shift and transit method to detect such alien worlds, as it is hard to pierce the glare of the host star that veils the planets around them. These techniques, along with sophisticated spectrographs and telescopes devoted to planet hunting, offer hope for finding more of them in the coming days, including a true twin for our planet Earth.

When a planet crosses directly in front of a star, it can dim the brightness of the star for a short period and can be caught in the field of view of telescopes. A similar effect can be seen when Earth passes the Sun, and if that journey is watched, a periodic dimming of sunlight occurs in exactly 365 days. Depending on the distance from the star, such periodic dimming of starlight can be considered a true sign of possible planets around that star. Once this has been confirmed, astronomers focus their telescopes to that location to glimpse the new planet, and it serves as a second layer of verification. Also, there is a gravitational tug-of-war between the host star and its planets. When the planet moves along the orbit around the star, the star wobbles. By measuring the wobbling effect, one can confirm the presence of the planet. These indirect detection methods are used to locate exoplanets because they, unlike the stars, are too faint to observe directly.

Planet hunting is an extremely demanding venture as opposed to mapping stars. Planets are buried in the cosmic ocean filled by waves of starlight. It requires a determined and collective effort to churn this ocean to discern hidden planets. It takes several years or even decades of observation and data analysis followed by confirmation with other ground- and space-based telescopes. While the estimated number of stars in our galaxy is around 200–400 billion, the planet count is still a few hundred – that alone speaks to the nature of challenge in the planet hunt.

Astronomers believe many small, rocky planets are in our cosmic neighborhood, but it is hard to locate them with our current capabilities. They estimate that planets similar to Earth could be orbiting one in every four stars like our Sun. If we generalize this as Earth-size planets – between one-half and two times the mass of Earth – we can easily predict that there are plenty of them in our galactic neighborhood. Maybe for every few stars, such planets exist. This means we don't need to look too far to find cousins for our own planet.

Detecting planets by looking at how they yank about their parent stars may not be very reliable, and the process is often associated with an element of ambiguity. We may not have the technology yet to image many of these newly found planets, as they are obscured by their host stars. But human imagination, more than machines, drives the planet hunt and is probably the most remarkable tool in our arsenal. Even ancient cultures have speculated that other solar systems might exist and that some would harbor other forms of life.

The possible abundance of Earth-like planets, though not a single one is confirmed yet, suggests that they are potential homes for life, where water along with other organic molecular matter might be present.

Again, we must think about our position in the universe. Is our planet a mundane world in a universe infected with life, or a special place fine-tuned to support life? If the recent discovery of hundreds of exoplanet candidates is of any indication, we may have to shed our weak anthropic view of the universe.

Ever since the day Galileo knocked on the doors of heaven with his newly found telescope to unveil the unknown, humans and their machines have evolved to become skilled planet hunters. Now, 400 years after the Galilean adventure, an array of space-and ground-based telescopes scans the sky with their unblinking eye for the exoplanets.

Among those machines, the Kepler observatory, bearing the name of Johannes Kepler (known for Kepler's laws of planetary motion) has an unparalleled position. Launched by NASA in 2009, Kepler is designed to search for Earth-like planets, small rocky planets in the habitable comfort zones from their host stars – neither too hot nor too cold.

Kepler scans about 150,000 stars simultaneously in a star-rich patch of the sky, considered to be the fertile ground for planets. When these planets cross the line-of-sight between the star and the observing eyes of Kepler, a dimming of the starlight is recorded by the sensitive instruments onboard the craft. An Earth-like planet will cause such a dimming of the parent star in about a year. This technique, known as a transit, is one of the methods employed in potential planet detection.

The Kepler observations have also identified about 700 stars believed to be harboring planets, with some of them even containing multiple planet systems. A conservative estimate puts at least 400 of them as capable of hosting planets with masses similar to that of Earth.

These potential planet-candidates need a second layer of verification before finding a place in the exoplanet catalog. To eliminate possible false alarms and identify the actual planets, the Kepler team has to follow-up with ground- and space-based telescopes. So far, Kepler has cataloged 2,740 planets candidates and confirmed 114 as planets – and it is hoped the count will grow. It discovered, for the first time, a planet with two Suns – a rare and unique finding that we could only imagine. A very recent malfunction (May 2013) of the Kepler observatory has cast doubt about its ability to further accelerate the hunt for much sought Earth-like planets. However, astronomers are optimistic that Kepler has gathered enough data that would enable them to sort out some astonishing findings even if the observatory does not recover from the failure.

Many believe that finding an Earth-like planet is not even a question of time. Rather, it is a question of having the right tools in the right place.

Detection techniques, such as transit and gravitational microlensing, will provide the basic details of alien planets. However, this is not an indication of any close contact with alien bodies, as the nearest exoplanet is at least a few light-years from us. Our fastest-moving spacecraft would take thousands of years to reach there for a direct observation.

So far, the majority of the known exoplanets are gas giants with extreme conditions, cold or hot, which leaves the position and structure of Earth, at least for time being, very special – a rocky planet in the habitable zone of its parent star.

The first ever detection of a rocky planet came in 2009, about 400 light-years away from our Solar System. But this planet was positioned too close to its parent star, with an average surface temperature of 1,000 °C, a disqualification to be considered a twin of Earth.

The confirmation of an Earth-like planet will indeed be met with much excitement, and that will spark a debate in the scientific community about our place in the universe. Kepler is also capable of picking up the fingerprints of life vis-a-vis the molecular evidence of water and oxygen. A rocky planet in a comfort zone is a perfect place for evolution to get under way – a breeding ground for life.

If our planet and lives are the result of a mere cosmic accident, it is bound to happen again in the vastness of space and time. On the other hand, if we are the unique outcome of a fine-tuned universe, we have the right to know the glory of that act.

After all, anthropocentric belief is not that appalling, if we regard ourselves as part of nature along with whatever we find. It makes sense to agree to the fact that the universe lives in us while we are part of it. We can only wish for harmony of the worlds.

Hunting exoplanets may be an innate reflection of our own self. Maybe we are trying to find ourselves or our own copies on those worlds. In Vedic mythology, in Indra's[6] palace in heaven there is a network of pearls called "Jewel net of Indra" that extends infinitely in all directions. The pearls are arranged in such a way that each is reflected in every other one throughout the whole net. This could be a mythological metaphor for various concepts, but one might wonder when we come across so many planets in the cosmos whether these planets are like networked pearls scattered across the fabric of the cosmos with a yet-unknown strand connecting them all. This strand connects the dust particles to the mighty stars, people to planets, and everything else. This broad interconnectivity rules the cosmos and possibly is the cause of the universe. Perhaps our dreams and imaginations about other worlds are just the desire of knowing this interconnectivity.

As history teaches us, in the sixteenth century, Earth was dethroned from its unique position in the universe, which ignited a new revolution in our thinking. Similarly, a world of many Earths will spark a debate about our place in the universe. Is our planet a mundane world in a universe infected with life, or a special place fine-tuned to support life? We don't know yet. Either way, exoplanets will continue to fascinate us. Our fastest-moving spacecraft will not take us there in the near future, but our imaginations will definitely do the job.

---

[6] Indra is the king of Gods in Hindu mythology

Figure 7.5. The other worlds. Dust disc indicates planets around a star. More exoplanets have been identified lately – the other worlds (Image credit: NASA).

REFERENCES

A Scientist Takes On Gravity – NYTimes.com (2013). A scientist takes on gravity – NYTimes.com. Available at: http://www.nytimes.com/2010/07/13/science/13gravity.html?pagewanted=all&_r=0. Accessed 2 Oct 2011.

BBC – Religions – Hinduism: Vishnu. BBC – Homepage. http://www.bbc.co.uk/religion/religions/hinduism/deities/vishnu.shtml. Accessed 18 Aug 2010.

Borges, J. L. (2007). The book of sand and Shakespeare's memory (Penguin classics). Edition. Penguin Classics.

Exoplanet Orbit Database|Exoplanet Data Explorer. (2013). Exoplanet orbit database|exoplanet data explorer. Available at: http://exoplanets.org/. Accessed 02 July 2013.

Gerlach, E., & Haghighipour, N. (2012). Can GJ 876 host four planets in resonance? *Celestial Mechanics, 113*, 35.

Haswell, C. A. (2010). *Transiting exoplanets*. Cambridge: Cambridge University Press. Edition.

Kashlinsky, A., et al. (2008). A measurement of large-scale peculiar velocities of clusters of galaxies: Results and cosmological implications. *The Astrophysical Journal. 686*, 49–52. Available at: http://iopscience.iop.org/1538-4357/686/2/L49/pdf/1538-4357_686_2_L49.pdf. Accessed 05 June 2010.

Kashlinsky, A., et al. (2009). A Measurement of large-scale peculiar velocities of clusters of galaxies: Technical details. *The Astrophysical Journal, 691*, 1479–1493. Available at: http://iopscience. iop.org/0004-637X/691/2/1479/pdf/0004-637X_691_2_1479.pdf. Accessed 30 June 2010.

Malfunction Could Mark the End of NASA's Kepler Mission – ScienceInsider. (2013). Available at: http://news.sciencemag.org/scienceinsider/2013/05/malfunction-could-mark-the-end-o.html. Accessed 16 July 2013.

Mandelbrot, B. B. (1982). *The fractal geometry of nature*. San Francisco: W.H. Freeman.

Mayor,M., & Queloz, D. (1995). A Jupiter-mass companion to a solar-type star. *Nature, 378*(6555), 355–359. Available at: http://www.nature.com/nature/journal/v378/n6555/abs/378355a0.html. Accessed 3 Dec 2009.

NASA – NASA and NSF-Funded Research Finds First Potentially Habitable Exoplanet . (2010). *NASA – NASA and NSF-funded research finds first potentially habitable exoplanet* . Available at: http://www.nasa.gov/topics/universe/features/gliese_581_feature.html. Accessed 13 Mar 2012.

On the origin of gravity and the laws of Newton – Springer. (2011). *On the origin of gravity and the laws of Newton – Springer*. Available at: http://link.springer.com/article/10.1007%2FJ HEP04%282011%29029. Accessed 23 Apr 2012.

Perryman, M. (2011). *The exoplanet handbook* (1st ed.). Cambridge: Cambridge University Press.

Seager, S. (2011). *Exoplanets* (Space science series). Tucson: University of Arizona Press. Edition.

The Project Gutenberg eBook of The Thoughts of The Emperor Marcus Aurelius Antoninus, by George Long. (2013). *The project Gutenberg eBook of the thoughts of the emperor Marcus Aurelius Antonius, by George Long*. Available at: http://www.gutenberg.org/ files/15877/15877-h/15877-h.htm#viii._52. Accessed 23 Mar 2012.

8

Holographic Universe: The Ultimate Illusion

There is a philosophy that says that if something is unobservable – unobservable in principle – it is not part of science. If there is no way to falsify or confirm a hypothesis, it belongs to the realm of metaphysical speculation, together with astrology and spiritualism. By that standard, most of the universe has no scientific reality – it's just a figment of our imaginations.

– Leonard Susskind (*The Black Hole War: My Battle with Stephen Hawking to Make the World Safe for Quantum Mechanics* 2008)

The mere thought that everything we know and love, including our own beings, is the result of information projected on a film like a 3D video may resonate with the familiar plots of science fiction movies. However, this is exactly what physicists stumbled upon while exploring the known limits of the universe. If this ultimate illusion is true, shockingly, we reside on a gigantic hologram and probably are little more advanced than the holographic images we create now.

This outlandish idea, known as the holographic principle, suggests that our 3D world is a profound illusion and is the pure materialization of natural laws painted on a distant cosmic landscape. Our daily experience and life itself are just some of the several ways by which this information takes shape in our world and we perceive it the way we do, calling it reality.

All the actions taking place in our universe, in the form of matter, energy, or even life, are the incidentals of information encoded on the cosmic edge. In other words, even we are the chosen characters of a holographic movie that unfolds as the cosmic wheel spins.

S. Mathew, *Essays on the Frontiers of Modern Astrophysics and Cosmology*, Springer Praxis Books, DOI 10.1007/978-3-319-01887-4_8, © Springer International Publishing Switzerland 2014

It may sound absurd to promote such an idea, but this is the natural extension of the vital knowledge we gained through the study of the cosmos and one of its most bizarre products – black holes. While it might cause jitters and shatter our general sense of well being, researchers like Craig Hogan, the director of Fermilab Center for Particle Astrophysics, supports this strange idea, encouraged by some recent experimental results, that "we are all living in a cosmic hologram."

We cannot avoid the discussion of black holes when it comes to the holographic nature of the universe.

THE BLACK HOLE WAR

About a decade ago, in an effort to save one of the pillars of fundamental physics, namely, the law of conservation of energy, the famous Dutch theoretical physicist and Nobel Prize winner, Gerard't Hooft, put forward a daring proposal (2000) on quantum gravity at the horizon of a black hole, which laid the foundations of the holographic principle.

Prior to this proposal, a prolonged war had been waged in theoretical physics without a clear resolution for decades, which even threatened one of the foundations of quantum physics. While studying the strange features of black holes, the well-known British physicist Stephen Hawking theoretically proved that black holes are not really black, in the sense that they emit radiation, now known as Hawking radiation. This emission could eventually destroy the black hole as it evaporates and disappears without any trace.

Hawking also asserted that this emitted radiation is featureless and devoid of any information. This essentially brings us to the conclusion that the complete evaporation of the black hole would erase all the past information permanently, including how it was formed in the first place, an idea that challenges one of the tenets of modern physics. Physicists may agree about the information scrambling but not on information erasing, as that is simply unacceptable.

We can always construct the past picture of an object or action from the current information, however long and hard the road may be. An analogy of this sort can be seen in our world and may emphasize the strength of this argument.

For instance, when an accident happens involving, say, vehicles, it is possible, though often hard, to pinpoint what happened just before the accident. Similarly, physicists believe that we can always make a past construct from the current information and that information is never lost completely. Think about how we could generate a picture of the Big Bang, even after 14 billion years, from the information that is a remnant of the event, though it was an extraordinarily hard task. To put it simply, permanent removal of information is impossible. "Thou shall not erase the information" seems like a commandment. Hawking's argument posed a threat to this decree.

Known as the black hole war, the echoes of this information paradox still reverberate in the scientific community. Leonard Susskind of Caltech on one side and Hawking on the other, a battle had been waging this war for decades, with neither of these modern-day warriors of physics willing to concede to a ceasefire. However, finally, Hawking agreed to the fact that black holes are information scramblers rather than information erasers.

Figure 8.1. A stellar black hole that belongs to a binary system as pictured in this artist's impression (Image credit: ESO).

Contrary to our expectations, physicists like paradoxes like the above one. The seemingly unsolvable paradoxes can lead us to magnificent new insights. When the black hole war was over, the ultimate winner was human intelligence and knowledge, not the warriors who led the intellectual attacks. It supplied a completely new way of thinking about how space and time could originate quantum mechanically. Space and time could emerge from the quantum convulsions as opposed to the pristine picture we had prior to the emergence of quantum mechanics. This led to a remarkable paradigm shift about space, time, matter, and bits of information about basically everything that we know now.

When the physicists found they could save the conservation of information tenet and data must be retrievable, it also provided the idea that this information must be recorded somewhere. Again, the best possible place to look for this was black holes.

No one knows what exactly happens inside the black hole, but everyone agrees that the area surrounding the black hole – the event horizon – provides the information about the black hole. In other words, we can generate a picture of the black hole by retrieving the information etched in the two-dimensional surface surrounding it. In fact, our so-called pictures of the black hole, what we see, are basically built on the information that surrounds it. Without this information, there is no black hole, neither the picture nor the imagination.

Now, Susskind and Gerard 'T Hooft extended this argument, refining the mathematics, to picture a cosmic black hole. Scientists have known for a long time that information plays a key role in the creation of a system. Our computers and robots are just metal and wires if no information is exchanged in the form of bits. Our brains are inanimate if no information is carried by the neurons. Our genes are futile if no information is available from the DNA, which instructs the body how to function.

In everything we know, information is key. Similarly, information about our universe must be encoded elsewhere. Like a hologram on our credit cards, which contains the information in a thin film and can generate 3D objects when viewed in proper light, the reality we are tempted to believe is actually just one way of viewing information printed on a distant cosmic film. What we see and experience as reality are shadows of the truth.

This is reminiscent of a metaphor in which the Greek philosopher Plato portrayed his views of human ignorance in *The Republic*, using an allegory called the parable of the cave. In the allegory, Plato compares humans to prisoners chained in a cave, unable to turn their heads. To the back of the prisoners, under the defense of the parapet, lie the puppeteers, who are casting the shadows on the wall in which the prisoners are perceiving reality. What the prisoners see and hear are shadows and echoes cast by objects that they do not really see. They mistake the appearance of objects for reality. They think the things they see on the wall (the shadows) are real, and they know nothing about the original objects. Those who escape the cave could go out and recognize the true reality, but they are treated as madmen by their fellow prisoners when they return and explain what they saw.

The world revealed by our senses is not the real world but only a pitiable copy of it, and the real world can only be comprehended intellectually. Perhaps this parable is meant to indicate humankind's struggle to know the truth through reasoning and progression. This is what the famous line from *The Republic* means, "To them, I said, the truth would be literally nothing but the shadows of the images."

The majesty and beauty of our universe is the projection of data that exists outside our accessible universe! How could this be true? We not only see and hear our world, but even experience it through other senses. We can touch the objects but not the holographic images, the reason we call them virtual. Listen to what researchers say. In the coming years, we should be able to interact with 3D holograms in real time. As we march forward, technology will have more and more movies and TV shows in 3D, and the cameras that generate them will get more compact and even fit into cell phones. At that point, we will be able to interact with photos, browse the web, and chat in entirely different ways that are considered off limits now. The 3D telepresence is closer than we imagine.

It has been reported (2008) that researchers from the University of Tokyo have developed 3D holograms that can be touched with bare hands. It has to be mentioned here that there are no laws of nature that prohibits such a scenario.

As our technology evolves with time it should be possible to generate our own 3D images that can interact with us. At this point, it is interesting to look at a philosophical argument suggested by the Oxford philosopher Nick Bostrom, which continues to attract a great deal of attention. He argued about the probability, though small, that we are living in a simulated environment created by an advanced civilization (2003) and are enjoying the conscious experiences in that simulation of what we call life. It also follows that as we evolve to a very advanced level, we might be able to program our future computers to create simulations that could experience their own versions of consciousness and life.

But, for physicists, that is a pure philosophical exercise. For them, it must be possible to gauge or at least get the clues of the true nature of the universe in a scientific fashion. As described earlier, the gravitational wave detectors are among the most sensitive scientific tools that can measure the ripples in the space-time fabric caused by the cosmic events such as star collapse.

One such detector, known as GEO 600, that can detect a fluctuation of an atomic radius over a distance from Earth to the Sun, recently recorded a noise that kept the researchers scratching their heads for a while. Craig Hogan of Fermilab thinks this is a vindication of what he suggested earlier (Hogan and Jackson 2009). According to him, the noise could be from the lowest possible units of space and time that can exist, called Planck scales. He suspects this fuzziness is emerging from the smallest units of the ingredients of the universe, and the universe is not continuous as Einstein thought but discrete units of space and time. The holographic noise the detector is receiving is an indication that on smaller scales space-time is quantized instead of smooth and continuous.

The grand picture of the universe that we see is the result of the quantum graininess that lies at the heart of reality. Imagine we zoom in on a picture, and as we get closer and closer, we will discover the pixels that make up the picture, and then it becomes fuzzy. The tiny discrete pixels generate a continuous and smooth picture for us on a larger scale. All the information we can conceive about that picture is encoded in the pixels, which are not continuous but distinct. Like a big picture emerges from the pixels, our universe is the culmination of bits of information that exist at the quantum level and so do we. Particles and energy are nothing but carriers of this information to create the universe.

As Shakespeare wrote in *The Tempest*, "We are such stuff as dreams are made on, and our little life is rounded with a sleep." (Act 4, Scene 1, 148–158)

At present, the holographic principle is in its hypothetical stage, motivated by black hole theory. But at Fermilab in Chicago, the scientists, led by Hogan, have devised a holographic interferometer to advance further observation. Perhaps it is just a stretch of wild imagination to make a statement that we are holograms, but it is not time to completely rule it out either. But the choice is clear, as some prisoners did in Plato's cave, that we should venture out and seek the truth about reality, or we will continue to say the shadow is real. Maybe we will find that our universe itself is embedded in a cosmic black hole and acting according to the information inscribed on its surface. Unless we get out, no light will reach us from beyond the event horizon.

Every physicist agrees that our universe is an interplay between matter and energy. But, what if our universe is just an information-exchanging network – an archetype proposed earlier by physicist David Bohm. A hologram contains all the information about a 3D image encoded in a 2D object only to be revealed under the right conditions. Every part of a hologram contains all the information possessed by the whole. What we experience here might itself be a holographic projection of processes that take place on a distant, 2D surface – a situation aptly described as "it from bit." In such a holographic universe, space and time are quantized, like pixels in a picture.

Our current picture of the universe is built on the information provided by photons, or light. It remains to be seen how gravitons will remake that grand cosmic picture. It is hard to reconcile the idea that gravity is the weakest of all known forces in nature, which has confounded physicists for decades. Now some theorists believe that gravity may be leaking into parallel universes or other dimensions, thus making it weaker in our universe. Such bizarre predictions are beyond any form of verification and thus remain outside the walls of scientific hypothesis.

Our imagination may be the predecessor of reality. But, do we have a limit on our imagination?

For many people, including scientists, it is a challenge to rethink the conventional definition of science, but if we want to take into account a much wider range of human experience, we will have to do so.

In order to make sense of a universe that could even be described as a hologram, we must look some of the bizarre objects of the universe. Here is a brief description of how scientists estimated the age of a black hole in 2010.

THE BIRTH DATE OF A BLACK HOLE

This has all the ingredients of an epic story. An awe-inspiring plot, the inconceivable vastness, the ultimate destruction followed by an unimaginable creation. It also offers a flashback. The action begins about 30 years ago, about 50 million light-years from Earth. A star, 20 times the mass of our Sun, goes into oblivion, flashing its death message to every corner of the cosmos. A Maryland schoolteacher, who happened to be an amateur astronomer, picked up that signal with a small telescope. Professional astronomers and their massive telescopes follow the event. The rest is told to us on November 15, 2010, when NASA, for the first time, announced the birth of a stellar black hole. The reports are that the baby black hole is feeding well on the cosmic materials and will thrive.

Ever since the theory of relativity predicted their existence and since John Wheeler gave them their name, black holes have become a household name. They represent the limit of human imagination. The active life of a star is a continuous battle of forces, gravity against nuclear force. The inevitable victory of gravity collapses the core of the star, and if the star is much more massive than the Sun, its destruction likely results in the birth of a black hole. Black holes come in different sizes and masses, with millions of them in our own galaxy, yet they present an ongoing enigma to scientists. Researchers haven't witnessed the birth of one, until now. But, this discovery is remarkable in many other ways, too.

First of all, the supernova explosion that marked the end of the parent star, known as SN1979C, was detected by former schoolteacher Gus Johnson in 1979, as it took place in a galaxy named M100, beyond our own. It turned out that the death of that star was an actual birth cry of a black hole, although it couldn't be confirmed at that time.

Usually the birth of a black hole is accompanied by a bright gamma ray burst (GRB). But SN1979C was a different type of supernova with no such intense gamma rays. Abraham Loeb, professor of astronomy and director of the Institute for Theory and Computation at Harvard University, who is one of the key figures in the latest findings, said: "It is very difficult to detect this type of black hole birth because decades of X-ray observations are needed to make the case." NASA's Chandra Finds Youngest Nearby Black Hole - ScienceNewsline (2010)

The ensuing observation was made possible by the Chandra X-ray space telescope along with a bunch of other telescopes. The consistent and steady emission of X-rays

Figure 8.2. The birth cry of a black hole. This composite image shows a supernova within the galaxy M100 that may contain the youngest known black hole in our cosmic neighborhood (Image credits: X-ray: NASA/CXC/SAO/D. Patnaude et al., Optical: ESO/VLT, Infrared: NASA/JPL/Caltech).

detected between 1995 and 2007 was considered as major evidence of a black hole. The Chandra space telescope, named after Subrahmanyan Chandrasekhar, who is famously known for the Chandrasekhar limit in astrophysics, is one of the major space telescopes operating in space along with Hubble. Coincidentally, October 19, 2010, was the one-hundredth anniversary of the birth of Subrahmanyan Chandrasekhar.

Is this black hole really 30 years old? It is 30 years old if our frame of reference is Earth, but it is 50 million years old based on the parent galaxy. But that is irrelevant for earthlings. We are separated from that galaxy not just by space but by time as well. Time, along with space, limits our view of the universe. An observer in the vicinity of the galaxy would say it is 50 million years old. Finally, from the point of view of the light coming from the supernova, no time has passed and no distance was crossed, so it is instantaneous.

All are right. But for us, it is a baby black hole, and it provides a rare opportunity to see how it develops from infancy and probably rules the cosmos 1 day.

Fine, we agree on the age of the black hole. What about the name? Astronomers don't name black holes. Their destiny is to be known by the event that created them. The reason is simple. We can't see them at any wavelength, and names are given to things that can be observed. We infer their presence by the evidence. Yes, this is a dilemma, because the universe is a story being scripted by the laws of physics, given to all but revealed to few.

Among those revelations black holes remain the most striking. Some see them as ultimate destroyers, but many perceive beauty in their actions.

THE BEAUTY OF THE BEAST

Too often, black holes are pictured as ultimate destroyers that suck up everything in their vicinity. In a dramatic shift that shatters the traditional wisdom surrounding these cosmic monsters, some physicists are giving them a new look, and if they are right our universe resides inside a black hole. Apparently, the beasts of the cosmos, long considered to be the synonym of death and destruction, are gaining some respect from the same laws of physics that created them.

The death of stars in the universe is somewhat similar to our own demise. No one seems to care about the poor and ordinary beings when they depart. But the deaths of the noble and privileged make an impact, and they are well celebrated. In a universe filled with more stars than humans on the planet, the death of small stars often goes unnoticed, and they may get a second life in the form of white dwarfs. An ordinary star like sun will have such a fate. The big ones, on the other hand, send their death message to every corner of the cosmos in an act known as a supernova. Furthermore their life after death, known as a black hole, is met with horror and fear.

The stars much more massive than our Sun are possible candidates to become black holes after they spend their active life. The life of a star is a constant battle between the force of gravity that tries to collapse it and the nuclear force that balances the onslaught of gravity. But, ultimately, nuclear fuel is exhausted and gravity wins – a black hole is born. Black holes come in different varieties, from microscopic in size to super massive ones found at the center of our Milky Way Galaxy, and in fact, in all galaxies.

Black holes represent the breakdown of the known laws of physics. When Einstein formulated the theory of relativity, he could not have imagined such a consequence for his theory. But, the seeds of destruction of space and time were already implanted in his famous laws that govern gravity. Initially, Einstein rejected such a possibility, claiming that nature would prohibit their formation. Ironically, it was his theory that predicted the very existence of black holes.

The direct detection of black holes is not possible, since light can't get away even from the event horizon. But astronomers have concrete evidence for the existence of black holes. They track the motion of stars around them and measure the secondary X-ray emissions from the hot gases to confirm and identify these cosmic holes.

In a recent paper, "Radial motion into an Einstein-Rosen bridge," Nikodem Poplawski of Indiana University (2010), reported that his theoretical study related to the motion of particles inside a black hole suggested that observed astrophysical black holes may be Einstein-Rosen bridges (wormholes), each with a new universe inside that formed simultaneously with the black hole. Accordingly, our own universe may be the interior of a black hole existing inside another universe.

Some researchers believe that our universe began with a singularity, where the known laws of the universe have no meaning. Similarly, it is assumed that inside a black hole the objects must propel towards the singularity that exists at the center of the black hole. But the motion of galaxies in the universe suggests that there is no preferred direction of motion in space, rejecting the notion that our universe is inside a black hole.

It remains to be seen how the new results explain such phenomena. At this point, Poplawski's results are speculative and are not going to change our conventional understanding of the universe. Yet, it is a completely different way of looking at these fascinating objects and the creation of the universe – an alternate reality.

Originally known as 'frozen stars,' black holes are the creation of the human imagination. The visionary physicist John Wheeler, a pioneer in nuclear fission who gave these stars the name of black holes, aptly wrote: "We live on an island surrounded by a sea of ignorance. As our island of knowledge grows, so does the shore of our ignorance."

Black holes continue to strain our imagination. It is estimated that our own galaxy has millions of them. The universe seems like an arena where the black holes perform their cosmic dance. On a grand cosmic scale, they could be the destroyers as well as creators. They might be harboring the secrets of the beginning. Yes, it is time to appreciate the beauty of these beasts that will bring knowledge.

Our discussion on black holes wouldn't be complete without more detail on NASA's Chandra X-ray observatory, which is specially designed to detect X-ray emission from very hot regions of the universe such as exploded stars and the matter around black holes.

THE STAR POWER OF CHANDRASEKHAR

In grandma's tales, we become stars in the sky when we depart our loved ones on this planet. This seems plausible in a universe where there are more stars than humans. But, what about a man whose life on Earth was devoted to the stars in the sky? He became, in a way, an even greater star gazer a few years after his death.

The Indian-American physicist Subramanyan Chandrasekhar devoted his Earthly life to studying stars and tried to know how they evolved and died – an unknown field in astrophysics in those days and only a few had delved into it at all prior to him. Even after his death in August 1995, it seems, he didn't want to give it up. The Chandra X-ray observatory, one of the major telescopes in space, is named after him. This is one of the flagship space observatories, and along with the Hubble Space Telescope it has provided a wealth of data and images that before then had existed only in the dreams of physicists.

Four years after Chandrasekhar's death, NASA launched Chandra X-ray observatory to an orbit about 86,500 miles above Earth. It is designed to detect X-rays from galaxies and matter that surround the black holes, to create a picture of the universe unknown to humankind and invisible to our naked eyes. At that altitude Chandra's eyes can reveal the objects that we can only imagine here on Earth, where the X-rays from such sources are absorbed by the atmosphere. The Chandra X-ray observatory is operated from the Smithsonian Astrophysical Center, Harvard University, Cambridge, MA. They process the

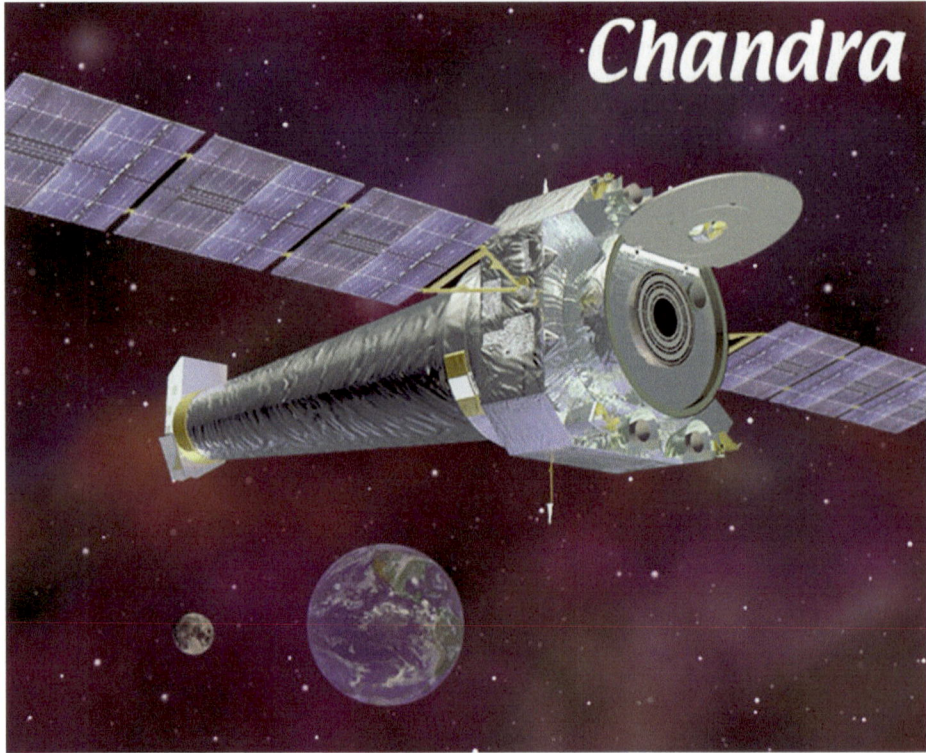

Figure 8.3. Chandra X-ray observatory (Image credit: NASA).

data and distribute it to scientists around the world for analysis. The center is an excellent resource for science education; their website (http://www.cfa.harvard.edu) offers more details.

Chandra is considered to be most sophisticated X-ray observatory to date and has revealed us an X-ray universe – a hitherto unknown universe that no other telescopes have ever ventured into. It imaged the event horizon of black holes, many spectacular nebulae and the glowing remains of exploded stars – and the list goes on.

Chandra has the unique ability to observe X-rays from particles just before they fall into the black hole. Its data has contributed much to the study of dark matter and dark energy. It also pointed to the existence of a massive black hole in the center of our Milky Way and found black holes across the universe. As its mission continues, Chandra will continue to reveal the unknown realms of our universe. Thus the legacy of Chandrasekhar lingers over the sky and in the hearts of people through the actions of this great observatory.

Chandrasekhar was born in Lahore, as his father was posted there as the Deputy Auditor General of the Northwestern Railways in British India. Later, he was transferred to Madras, where Chandrasekhar completed his Bachelor's degree in physics before moving to Cambridge on a scholarship. In 1936, Chandra married Lalitha Doraiswamy, with whom he had attended physics classes at Presidency College in Madras. Chandra was the nephew of Nobel laureate C. V. Raman, the first Asian to receive that highest honor in science.

In his early twenties, when Chandrasekhar crossed the ocean aboard a ship to study at Cambridge in the UK, only a few thought that this young Tamil would one day lead us through the cosmic ocean, which was mostly unexplored and unexplained at that time. During his time at Cambridge, he had the opportunity to meet the physics stalwarts of that era, namely Dirac and Bohr. Subsequently, he was awarded a doctorate in physics in 1933.

It's been said that during his voyage from India to England, Chandrasekhar figured out the upper limit of mass a star can have to become a white dwarf at the end of its life. This maximum mass, later known as the Chandrasekhar limit (1.44 solar mass), is one of the foundational aspects of modern astronomy. Any star with masses more than this will end up as a neutron star or black hole following their violent death, after shining for billions of years. The stars like our Sun, after another 5 billion years of its life, will end up as a white dwarf, deprived of its current brightness and energy that allow us to survive now here on this planet.

In 1937, Chandra joined the faculty of the University of Chicago and remained there for the rest of his life. He was elected a Fellow of the Royal Society of London in 1944 and was the sole editor of *The Astrophysical Journal* from 1952 to 1971. Among the numerous other prizes and honors, Chandra was awarded the Nobel Prize in Physics in 1983 for his theoretical studies of the processes related to the structure and evolution of stars.

Chandra, as he was known among students and colleagues, was a popular teacher as well. He helped a number of students to complete their doctoral thesis. His research explored nearly all branches of theoretical astrophysics, and he published many books, each covering a different topic, including one on the relationship between art and science. "Chandra probably thought longer and deeper about our universe than anyone since Einstein," wrote Martin Rees, English cosmologist and Astronomer Royal.

In addition to his scientific endeavors, Chandrasekhar is also well known for his admiration for the philosophy of science and his interest in music and literature. In his book, *Truth and Beauty: Aesthetics and Motivations in Science,* Chandrasekhar explored how artistic and scientific creativity, though appearing very different, are really similar in many aspects.

So, going back to our question of the illusive universe. Are we in a position to proclaim the universe is an illusion or not? By the time we reach the answer of that question, we may not even be able to distinguish between reality and illusion. However, in the meantime, the grand universe will continue its cosmic dance.

DOES THE MATERIAL WORLD EXIST?

The holographic description of the universe opens up the question about the existence of material world. If we need to know about something, we need information; this is fundamental to our physical universe. Information reaches us in the form of particles or waves, which manifest as energy or matter. We as conscious beings participate in the process of creating the physical universe. This may sound extremely philosophical; however, many physicists are inclined to entertain such seemingly strange ideas.

In his book *Decoding Reality: The Universe as Quantum Information* (2010), physicist Vlatko Vedral of Oxford University discusses how to view all natural and physical phenomena using information as the fundamental building block of reality. He writes, "In biology, for example, an event could be a genetic modification stimulated by the environment. In economics, on the other hand, an event could be a fall in a share price. In quantum physics, an event could be the emission of light by a laser when it is switched on. No matter what the event is, you can apply information theory to it. That is why I will be able to argue that information underlies every process we see in nature."

The Latin phrase *creatio ex nihilo* stand out as the dogmatic principle that many depend on to explain the universe, especially religions. However, that requires a creator, and we do not know how the creator is created. The information theory, as advocated by Vedral, suggests that information can created out of nothing and the information creates everything.

Now, if the cause of the universe is information, in the absence of that information the universe should not exist. What about us? It can be argued that even our own genetic code is information from which we unfolded.

According to French mathematician philosopher René Descartes the only evidence for existence is *cogito ergo sum,* which means "I think, therefore I am." In his work *Principles of Philosophy* (1644), he argued that the cause of our existence is the thought. This philosophical view is widely popular and considered the foundation of modern Western philosophy. This statement does not necessarily imply the existence of the human body but more accurately a description of the existence of the mind.

Philosophical arguments alone cannot answer our questions about the existence of our universe or our own self. Yet there was always an overlap between deeper questions in physics and philosophy. Physicists who ponder the most fundamental questions always wondered about the nature of reality.

As mentioned earlier David Bohm is one of the physicists who delved into this subject more than any other well-known physicists. His radical views on quantum mechanics attempted to answer the questions linked to matter and consciousness. He criticized the fragmented approach in science or the downward-looking reductionism to search for the truth. He wrote "The notion that all these fragments are separately existent is evidently an illusion, and this illusion cannot do other than lead to endless conflict and confusion. Indeed, the attempt to live according to the notion that the fragments are really separate is, in essence, what has led to the growing series of extremely urgent crises that is confronting us today." (2002)

According to Bohm, nothing is actually separate from anything else. He believed in an "unbroken wholeness" and proposed a theory with a new paradigm. He termed it the implicate order, in which not only the apparent distinctiveness of subatomic particles is illusory but they are misleading in our effort to know the ultimate reality. He also advocated a strong interconnectedness between mind and matter, as he believed they constantly change in relation to one another.

Bohm grappled with the existing notions of fragmentation and laced spirituality with his scientific pursuit. On a philosophical level, it is uplifting to accept the idea of

Figure 8.4. Is the universe a hologram or a figment of imagination? (Image courtesy of Flickr Creative Commons. Used with permission).

wholeness and it is comforting to think along those lines. However, the universal inter-connectedness and the cosmic consciousness have a long way to go to gain some sort of serious consideration in the scientific world.

When quantum mechanics was formulated, it wasn't much concerned about human consciousness or other topics related to psychic phenomena, such as telepathy and out-of-body experiences. However, later on, different interpretations in distinctive scientific fields created many different terms, including quantum consciousness. The tendency to relate these subjects to quantum mechanics is somewhat imprudent. Ultimately, the predictions must be tested to meet the scientific standards.

We should also recognize that the concept of the oneness of all beings was not an original notion of modern physics. A similar concept had appeared in early Greek philosophy and also in the Upanishads. The Isa Upanishad says:

He who perceives all beings as the Self for him how can there be delusion or grief, when he sees this oneness (everywhere).

REFERENCES

A Penn State electronic Classics Series publication: http://www2.hn.psu.edu/faculty/jmanis/upani-shads/upanishads1.pdf. Accessed 25 June 2013.

Barbour, I. G. (1974). *Myths, models and paradigms*. New York: Harper & Row.

Bohm, D. J. (1952). A suggested interpretation of the quantum theory in terms of 'Hidden' Variables I, II. *Physical Review, 85*, 166–193.

Bohm, D. J. (1953). Reply to a criticism of the causal re-interpretation of quantum theory. *Physical Review, 85*, 389–390.

Bohm, D. (2002). *Wholeness and the implicate order* (Reissue edn.). Routledge.

Bohm, D., & Peat, F. D. (1987). *Science, order, and creativity*. New York: Bantam.

Bostrom, N. (2003). Are you living in a computer simulation? *Philosophical Quarterly, 53*, 243–255.

Briggs, J. P., & Peat, F. D. (1984). *Looking glass universe: The emerging science of wholeness*. New York: Simon & Schuster.

Dirac, P. A. (1942). The physical interpretation of quantum mechanics. *Proceedings of the Royal Society A, 180*, 1–40.

Einstein, A., & Rosen, N. (1935). *Physical Review, 48*, 73.

Fuller, R. W., & Wheeler, J. A. (1962). *Physical Review, 128*, 919.

Goswami, A. (1995). *The self-aware universe*. Tarcher.

Haisch, B. (2009). *God theory, the: Universes, zero-point fields, and what's behind it all*. Weiser Books.

Hatfield, G. (2002). *Routledge philosophy guidebook to descartes and the meditations (Routledge Philosophy GuideBooks)*. Routledge.

Hogan, C. J., & Jackson, M. G. (2009). Holographic geometry and noise in matrix theory. *Physical Review D, 79*(12), 8.

Iwamoto, T., Tatezono, M., Hoshi, T., & Shinoda, H. (2008, August). Airborne ultrasound tactile display. SIGGRAPH 2008 New Tech Demos.

James, A., & Rees, D. A. (1963). *The republic of Plato* (2nd ed.). Cambridge: Cambridge University Press.

Krishnamurti, J. (1985). *The ending of time (Dialogue)* (1st edn.). Harper & Row.

Lanza, R. (2010). *Biocentrism: How life and consciousness are the keys to understanding the true nature of the universe* (1st edn.). BenBella Books.

NASA's Chandra Finds Youngest Nearby Black Hole - ScienceNewsline. (2010). NASA's Chandra finds youngest nearby black hole - ScienceNewsline. [ONLINE]. Available at http://www.science-newsline.com/articles/2010111602380000.html. Accessed 25 Oct 2012.

Nikodem, P. (2010). Radial motion into an Einstein-Rosen bridge. arXiv:0902.1994v3 [gr-qc] 13 Apr 2010 *Physics Letters B, 687*(2–3), 110–113.

Peter, A. (1994). *Creation revisited.* Harmondsworth: Penguin.

Radin, D. (2009). *The conscious universe: The scientific truth of psychic phenomena* (Reprint edn.). HarperOne.

Stenger, V. J. (1995). *The unconscious quantum* (1st edn.). New York: Prometheus Books.

Stenger, V. J. (2012). Available at: http://www.colorado.edu/philosophy/vstenger/Quantum/QuantumConsciousness.pdf. Accessed 25 Aug 2012.

Susskind, L. (1995). The world as a hologram. *Journal of Mathematical Physics, 36*(11), 6377–6396.

Susskind, L. (2008). *The black hole war: My battle with Stephen Hawking to make the world safe for quantum mechanics.* New York: Little, Brown.

Susskind, L. (2009). *The black hole war: My battle with Stephen Hawking to make the world safe for quantum mechanics* (Reprint edn.). Back Bay Books.

'T Hooft, G. (2000). The holographic principle. In Z. Antonino (Ed.), *Basics and highlights in fundamental physics* (pp. 72–100). Erice, Sicily: European Physical Society.

Talbot, M. (1993). *Mysticism and the new physics (Compass)* (Revised ed.). Penguin Books.

Talbot, M. (2011). *The holographic universe: The revolutionary theory of reality* (Reprint edn.). Harper Perennial.

Talbot, M., & McTaggart, L. (2011). *The holographic universe: The revolutionary theory of reality.* New York: Harper Perennial.

Vedral, V. (2010). *Decoding reality: The universe as quantum information* (First Thusth ed.). USA: Oxford University Press.

von Neumann, J. (1958). *The computer and the brain.* New Haven: Yale University Press.

von Neumann, J., & Burks, A. (1966). *Theory of self-reproducing automata.* Urbana: University of Illinois Press.

Youngest-Ever Nearby Black Hole Discovered – NASA Science. (2011). Available at: http://science.nasa.gov/science-news/science-at-nasa/2010/14nov_babyblackhole/. Accessed 13 Feb 2011.

9

Mind over Matter

Nature has no reverence towards life. Nature treats life as though it were the most valueless thing in the world.... Nature does not act by purposes.

– Erwin Schrödinger, Austrian-born physicist and theoretical biologist who was one of the fathers of quantum mechanics. (2012)

It may be argued, scientifically, that matter originated first, and the interactions of networked matter created everything else, including life, mind, and consciousness. But, a deeper analysis suggests, as some scientists believe, that consciousness is elemental and everything else is secondary. Is the mind the matrix of all matter?

Our scientific quest, based on reductionism, tempts us into believing that fundamental particles are really fundamental. They create our universe at a basic level, as they do our bodies and brains. But, if the brain is just a collection of particles that constitute cells, we wouldn't be much different than a rock or a piece of wood.

OUR NEURAL NETWORKS

The cells that make up the brain interact within a complex network that communicates among various segments of the brain. This exchange of information gives rise to non-physical attributes, such as mind, consciousness, and, as some believe, a soul within the human body. In the absence of such information trading, even when the brain is physically present, the meaning of existence is lost.

S. Mathew, *Essays on the Frontiers of Modern Astrophysics and Cosmology*, Springer Praxis Books, 133
DOI 10.1007/978-3-319-01887-4_9, © Springer International Publishing Switzerland 2014

It has been estimated that the human brain contains about 10^{11} neurons and 10^{15} synapses,[1] which could in theory form a humongous number distinct connectivity patterns. However, brain circuitry is neither random nor regular, and the information content of a single human brain remains far greater than the number of fundamental particles in the whole universe, let alone just the complete biochemical specification of that individual brain (2009). Thus network connectivity is necessarily more informative than the entire molecular profile of each of all of its neurons, including the expression of every gene and protein constituting the biophysical machinery at the basis of neuronal electrophysiology.

This neural circuit is responsible for producing all the different human faculties such as perception, thought, and memory. Research institutes such as the Max Planck Florida Institute study the complex synaptic networks of the brain, which hold the key to understanding who we are, why we behave the way we do, and how the debilitating effects of neurological and psychiatric disorders can be ameliorated.

Understanding the functional and structural features of neural circuits requires a collective effort from researchers with different expertise. The research and developments in this field show us how different disciplines come together that were once thought to be completely detached. The mind-matter relationship has come to the forefront of twenty-first century research.

The earlier scientific thinking about the mind-matter relationship was appealing, as it seemed lucid and coherent and it is still held by many modern researchers. It's simple and plain – matter (elementary particles, atoms, and cells, in that hierarchical order) came first, followed by life and consciousness. Accordingly, consciousness is an emergent property of the interactions of fundamental particles. The tangible creates the intangible. The tangible matter generates the intangible and unbounded entity of the mind, which is considered the ultimate source of our inner voices.

Many researchers (Ascoli 2013) believe that the relationship between mind and matter has perhaps been, in one form or another, the most debated issue in the history of human thought, and it still constitutes, in the modern "mind-brain" incarnation, an open scientific and philosophical problem. Specifically, why do certain brain states "feel" like something and why do specific brain states feel the way they do?

Some believe that the potential solution for this problem can come from mathematics. There are proposals that advocate a mathematical solution of the relationship between brain and mind that is consistent with contemporary philosophical positions (Feinberg 2012). Tremendous advancements in physics, chemistry, and biology provided an increasingly unified understanding of the material world. Bridging neuroscience with a quantitative description of inner subjective life may provide a fundamental closure to human scientific inquiry.

In an article entitled "The Mind-Brain Relationship as a Mathematical Problem," Giorgio A. Ascoli of Krasnow Institute for Advanced Study, George Mason University, writes, "a satisfactory answer can ultimately come from mathematics, if the abstract spaces of brain activity and mental content can be quantitatively characterized and geometrically

---

[1] Synapses are connections between neurons through which information flows from one neuron to another.

Figure 9.1. The cosmic connection. To completely understand the brain is equivalent to knowing the whole picture of the universe (Universe-CC Image courtesy of Chris Tolworthy on Flickr. Used with permission).

mapped onto each other. Such a 'solution' will connect the conscious mind and the relevant aspect of brain states by demonstrating the equivalence of their properties, much like statistical mechanics and thermodynamics are nowadays accepted as one and the same phenomenon, even if they are practically treated as distinct for every day purposes."

While everyone agrees that consciousness defines existence and reality, the mechanism that generates this phenomenon is controversial and mostly unknown. In the classical approach, a trigger in the form of an electrical pulse in the brain caused by an external stimulus originates at one location, and that information propagates to various cells, somewhat like a domino effect, yet it remains local in nature. The current neurophysio-logic understanding of consciousness asserts that it is a manifestation of the emergent firing of neurons, the nerve cells in the brain.

Thus, if a sufficiently complex system can generate consciousness, why can't machines mimic the same? Computers that execute algorithms, at some point, must be able to generate consciousness, because it is the result of complex networks performing information exchange among different components.

Our machines, such as computers, can store, process, and analyze information. And, some advanced algorithms, such as artificial intelligence, can imitate some other features, like muscle movement, pattern recognition, or even the sense of taste, smell, etc. What stops them from creating other attributes of consciousness and self-awareness? What about emotions and feelings? Is the brain an interface that bridges a super consciousness with human activities or is it a pure source of consciousness?

It is generally agreed that all the laws of physics are computable. In other words, a complex machine can simulate the underpinnings of the brain at some point. If the laws of nature are computable, the workings of the brain and consciousness also must be comput-able. But, researchers have no clue how to incorporate such laws, nor are they confident that our current algorithms are capable of doing that. That's where the findings of Oxford physicist Roger Penrose become revolutionary (1994).

Penrose argued, to the dismay of many researchers, that our current science is inca-pable of creating artificial consciousness. Even if we can amass unimaginable computing powers, consciousness will still remain a fantasy. There is some missing link. What could be that unknown piece?

Penrose sought a deep and initially somewhat vague connection between conscious-ness and quantum physics. His groundbreaking work came to the public domain through his popular book *The Emperor's New Mind* (2002). He suggested that brain cells seem to execute quantum mechanical antics rather than the classical tricks to generate conscious-ness. Precisely, consciousness is created by some mysterious quantum mechanical phenom-enon that takes place in brain cells. And, unless we master that technique, consciousness will remain a mystery.

Working along with Stuart Hameroff of Arizona University, Penrose identified the cellular components that perform this quantum mechanical mishmash, or nature's quantum computers (1996). It turned out to be long and thin hollow tubes of protein about ten-millionths of an inch in diameter that network the cellular structure, known as microtubules.

Identifying the physical root of consciousness isn't enough to know the mechanism by which these tubes perform quantum mechanical computations. Quantum theory describes the underpinnings of matter and energy at the most fundamental level. Accordingly, an object is a "wave of possibilities" and it exists as a superposition[2] of all its possible quantum states simultaneously. Similarly, if we consider the brain as a quantum system, when the brain attempts to solve a problem, billions of different choices exist simultaneously as available outcomes. However, the numerous wave functions collapse

---

[2] In quantum mechanics a particle or object can occupy all of its possible quantum states simultaneously. That's the meaning of superposition.

and return a single choice from all the possible cellular traffic; this one becomes the conscious thought.

Also, in quantum behavior non-locality is the key. This essentially means the activity at one place affects the activity in another part of the brain without going through a domino effect needed by classical behavior. The information is entangled everywhere rather than being transported from one place to the other. Perhaps quantum information is hardwired to the entire universe. It's always been there in the universe waiting to be shared. Our brains are connected to the entire universe – a daring thought, some believe, that might 1 day explain paranormal effects such as telepathy or remote viewing.

The scientific attempts to unravel the workings of the brain and consciousness have a long way to go even if the quantum mechanical behavior proposed by Penrose is proved right. Recent research indicates that our sense of smell, photosynthesis, and even the navigation of birds all depend on strange quantum effects (FQXi Community 2013).

WHICH CAME FIRST, MIND OR MATTER?

The issue of mind-matter origin is passionately debated in scientific circles and will continue to be for a long time. However, the concept of consciousness is weaved into the fabric of Hindu philosophy. In Hindu thought, consciousness has an independent existence and is part of the ultimate reality, not an emergent phenomenon as conceived by science. Our individual minds borrow consciousness from the omnipresent Brahman, which describes the reality as sat (truth or eternal), cit (pure consciousness), and ananda (ideal bliss). Existence is a universal reality that transcends the manifested universe.

The fundamental particles, considered to be the building blocks, are inseparable from the energy field that pervades the universe. Their existence may manifest as energy or matter, but the state of existence is eternal in nature. The bewildering activity our brain performs to create consciousness is neither local nor totally new. Our brain is entangled with the entire universe and preexisting consciousness. It's the intangible that creates the tangible. We do not have evidence, as of now, to establish such a principle, but it's something researchers can't ignore as they nail down the science of consciousness.

Sri Aurobindo, the great Indian philosopher and poet, introduced the evolution idea into Vedantic thought through an unfamiliar and unmapped level of consciousness (Aurobindo 1972): "Evolution is not finished; reason is neither the last word nor the reasoning animal the supreme figure of Nature. As man emerged out of the animal, so out of man the superman emerges." He called this the Supermind. Our individual minds and bodies are parts of this principle, and it is present in the Satchidananda. The Supermind generates mind followed by life and matter. Rejecting the idea of renunciation of the material world, proposed by some schools of thought, he suggested it is possible to transcend and transform human nature and evolve spiritually, not just materialistically.

Aurobindo wrote, "Be conscious first of thyself within, then think and act. All living thought is a world in preparation; all real act is a thought manifested. The material world exists because an Idea began to play in divine self-consciousness."

Some say atoms and fundamental particles make up the universe, others that strings of vibrating energy are the building blocks, and a few suggest the material world is a grand illusion. The list of building blocks grows forever, yet none of these answers is complete.

Where should we look? Is the entire universe nothing more than God's dream, as some idealists like to believe, or is it an absolute entity independent of our practices and beliefs?

How small can we go? The fifth century BC Greek philosopher and scholar Democritus conjectured that the universe consists of empty space, and an (almost) infinite number of invisible particles that differ from each other in form, position, and arrangement. He postulated that all matter is made of indivisible particles called atoms ("atom" in Greek means indivisible): "Nothing exists except atoms and empty space; everything else is opinion. The worlds are unlimited. They come into being and perish. Nothing can come into being from that which is not, nor pass away into that which is not. Further, the atoms are unlimited in size and number, and they are borne along in the whole universe in a vortex, and thereby generate all composite things – fire, water, air, earth."

Though modern science adopted the name atom as conceived originally by Democritus, it is not indivisible any more. Scientists identified protons, electrons, and neutrons as the constituents of the atom in the nineteenth and twentieth centuries. Furthermore, protons and neutrons are found to be composite particles of quarks and gluons that bind them together.

According to Sri Aurobindo, the whole is the indivisible One (Self-Conscious Being, Sat), the parts are the Many. Both are partial truths. The Absolute is the infinite potential, which is greater than either the whole or the parts but contains and exceeds both, since it also includes that which is beyond manifestation.

Whereas Europeans later revived the atomism conceived by Greek philosophers, ancient India envisioned almost the same idea. In Sanskrit, *paramanu* has been translated as the smallest entity that cannot be divided further, while *anu* is considered to be atoms. Mysteriously, the Vedas and their interpretations do not offer many details about this atomic view, and for that reason atomism was not projected as a prominent thought in the Hindu school of philosophy. It has been generally agreed that the Indian and Greek versions of atomism developed independently in the fifth or sixth century B.C.

The idea of atomism, however, flourished in Buddhist philosophy through logical presuppositions. Early Buddhist thoughts about the world had as their foundations the Hindu elements of air, earth, fire, water, and ether. Ether is often referred to as the void. It plays an important role in Buddhist thought. The Hridaya sutra, which belongs to the Mahayana text, repeatedly says, "Form is emptiness, emptiness is form." When one looks closely at the structure of the atoms, it is easy to comprehend how an atom springs from predominantly empty space that separates the nucleus and electron cloud that surrounds it.

The internal structure of the atom, as revealed by scientific pursuit, is strange enough to pose many profound questions, such as the nature of reality. One might think of an atom as a solid ball with neutrons and protons in the nucleus and further small electrons zipping around them. Based on scientific analysis, this is not even close to the truth. Atoms are mostly empty, to be precise, 99.999999 % empty. The emptiness, the space between the nucleus and electrons, is what makes up the lion's share and what makes an atom an atom.

However, the meaning of emptiness described in Buddhism is markedly different from nothingness. People usually relate nihilism with Buddhism. But the idea of emptiness in Buddhism does not reject ultimate reality, as it proclaims emptiness as a form and form as emptiness. Emptiness, in Buddhism, is neither non-existence nor does it promote non-reality as in nihilism.

Later on, mostly in the twentieth century, physicists exploring the fundamental world have identified 12 building blocks that are the essential to the makeup of matter. Scientists also categorized four elementary types of forces acting among these particles – strong, weak, electromagnetic, and gravitational force. The now-famous standard model of particles suggests that our everyday world, including our own bodies, is made up of these elementary particles and the forces that interact among these particles.

In recent times, the much-celebrated string theory promised an elegant picture of the universe with vibrating strings of energy as the underlying entity that makes up everything, including these fundamental particles. This radical idea, however, still remains in the realm of philosophy, as it has failed to provide any experimental or observational evidence thus far. However, the unparalleled similar views that guided ancient wisdom and modern science as they sought the answers to the most fundamental entity are worth exploring.

The apparent solidity and form – the qualities attributed to all objects – are, in fact, an illusion on a microscopic scale. The atoms that make up objects are inherently empty, although these very same atoms make up elements, and they in turn create organic and non-organic compounds. Even 99 % of the mass of the human body is made up of just six elements – oxygen, carbon, hydrogen, nitrogen, calcium, and phosphorus – and their constituents are atoms.

The model "All in One and One in All" has its foundation in quantum mechanics. Schrodinger in his book *What is Life* (1961) remarked, "Consciousness is never experienced in the plural, only in the singular. How does the idea of plurality (emphatically opposed by the Upanishad writers) arise at all?….[T]he only possible alternative is simply to keep the immediate experience that consciousness is a singular of which the plural is unknown; that there *is* only one thing and that what seems to be a plurality is merely a series of different aspects of this one thing produced by deception (maya) – in much the same way Gaurisankar and Mt. Everest turn out to be the same peak seen from different valleys."

At the most fundamental level, the particles and forces may converge, and most probably, we haven't yet mastered the laws of nature that dictate this interplay. Until then, the illusion of form and shape instill in us the realities we claim to be true.

Even according to quantum physicists, particles are not what they seem. They are merely the tendency to exist rather than material objects. They are not the tiny spherical balls that could bounce off each other, as one might assume. The emptiness or vacuum that we portray as the absence of any material is not empty in that sense. Virtual particles always pop in and out of existence in the vacuum. Particles and forces all are interconnected and interdependent, and they are the cause of everything that is manifested and 'unmanifested.'

Perhaps everything that originated and existed is dependent on a single underlying entity that has neither yet been found by science nor revealed to us by ancient wisdom, at least in our own language. Even consciousness is part of that scheme, in addition to the normal senses of touch, sight, hearing, smell, and taste.

The philosophy of emptiness reflects a big paradox about the true nature of the world. We know that matter and energy are different manifestations of the same property and can be interchanged. The ancient Vedas, in their own language, describes the unmanifested.

With the curiosity of a schoolchild, one could ask what makes up both the existent and the non-existent? At this point, science, which explores mainly the world of the manifested, cannot provide a final answer. The principle of dependent origination denies the existence of anything with an independent or intrinsic identity. In that sense, emptiness is as expressive as material objects. Plurality and inclusiveness, the key elements of Eastern philosophy, have a broader appeal to philosophers and scientists alike.

It has been estimated that our observable universe contains about 10^{80} atoms. But it is also estimated that the universe we describe materially is just about 5 % of the known universe!

Our philosophers and scientists must look at reality without any prejudice of views, ideas, and perceptions. As Edgar Allen Poe wrote in 1845, "Deep into that darkness peering, long I stood there, wondering, fearing, doubting, dreaming dreams no mortal ever dared to dream before."

The recent experimental confirmation of the Higgs boson teaches us that the advancement of science is the result of doubting and dreaming of the mind. Unless we continue to do so we won't be able to know whether men are the dreams of Gods or Gods the dream of men.

HIGGS BOSON MANIFESTO

The announcement on July 4, 2012, about the Higgs boson discovery by the European Organization for Nuclear Research (CERN) set off a fireworks celebration of sorts among the scientific community. Although the Higgs boson is not the last word in our search for answers to our mysterious universe, it holds the potential to advance our understanding of the genetic code of our cosmos.

The Higgs boson with its metaphorical name, the God particle, has long attracted the attention of pundits and rookies, priests and philosophers, and believers and atheists alike. Once again, the debate over the nature of fundamental reality and consciousness is in the forefront. When the physicist Leon Lederman coined the term "God particle," he was not alluding to a literal God (personal or impersonal), even though the particle is "so central to the state of physics today, so crucial to our final understanding of the structure of matter, yet so elusive." Rather, he said, it was because "the publisher wouldn't let us call it the Goddamn Particle, though that might be a more appropriate title, given its villainous nature and the expense it is costing."

The elusive nature of the particle, along with its popular name, created a niche in the public psyche. It's commonly connected to the Big Bang and to the creation of the universe, or more generally, the creation of something out of nothing.

Normal matter, which creates stars, planets and even life, is made possible by fundamental particles called fermions (like electrons, the basic unit of charge) and bosons (like photons, the basic unit of light). In fact, we manipulate these particles in our everyday life so much that they have been employed to control everything from electronic circuits to solar cells. The Higgs boson is just one of the members of that family called bosons, along with photons and gluons.

STANDARD MODEL

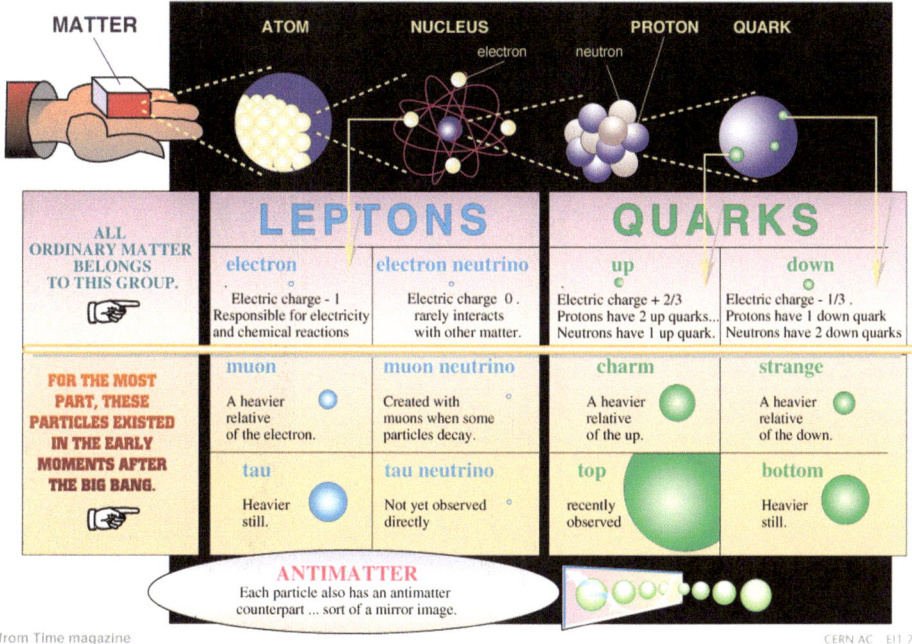

Figure 9.2. Constituents of matter (Image courtesy of CERN).

According to the standard model of particle physics, the Higgs boson is the manifestation of an invisible force field known as the Higgs field that permeates our cosmos. In the early moments of the Big Bang, particles zipped through the available universe and were massless. Then the Higgs field dragged the particles and they felt mass for the first time, according to the Higgs mechanism.

The theoretical framework that describes the interactions between elementary building blocks (quarks and leptons) and the force carriers (bosons) is known as the standard model.

What kind of interaction generates mass? This age-old question demanded a scientific explanation. The physicist Peter Higgs proposed a clever and elegant solution to this problem nearly half a century ago. In his theoretical model, the particle mass arises from complex interactions they have with the pervading Higgs field. Different particles interact with the Higgs field with different strengths, making them heavier or lighter.

In the absence of this invisible Higgs field, all the fundamental particles in the universe would zoom through the universe like shadows of reality. There would be no atoms, molecules or, for that matter, no life.

While the name 'Higgs' in the Higgs boson is commonly attributed to Peter Higgs, who is well known, the name boson has its roots in Satyendra Nath Bose, who is often overlooked. Bosons are virtually everywhere, and we cannot escape from these force-carrying particles of nature that enable inanimate matter to interact with each other to create a lively universe.

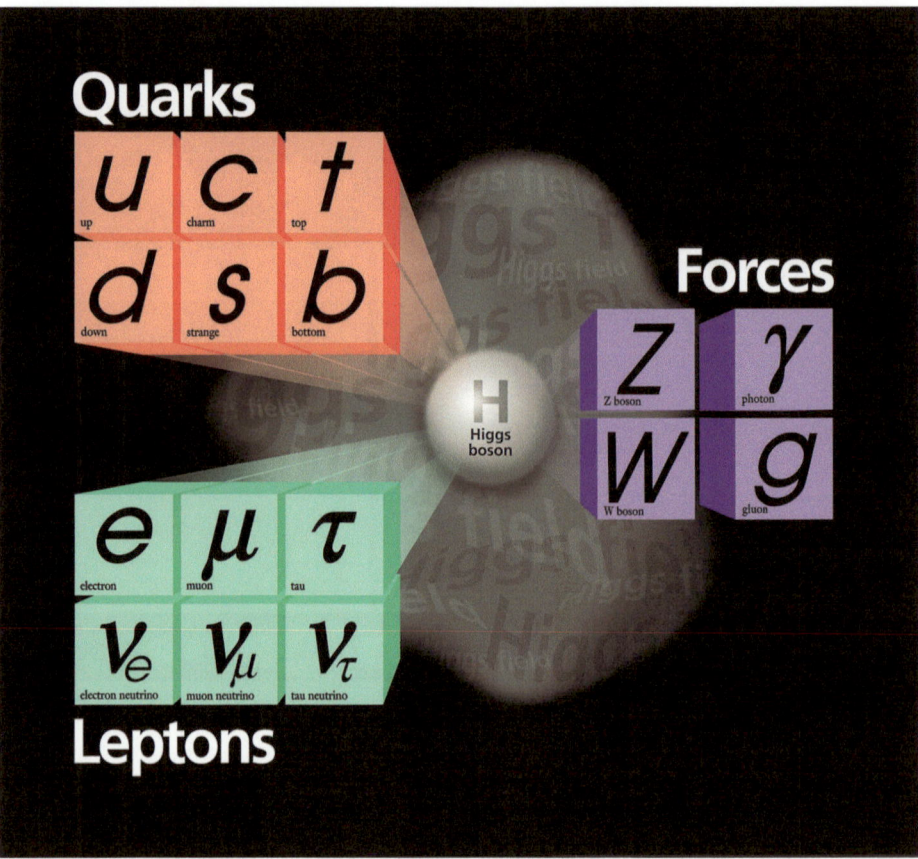

Figure 9.3. The standard model of particles with the newly discovered Higgs boson (Image courtesy of Fermilab).

It has been assumed for a long time that the Higgs bosons are responsible for providing mass to everything in the universe. Now, you can imagine why physicists have been seeking such a particle that was born out of pure imagination and mathematical equations. Specifically, the Higgs boson emerged from Higgs' vision and theoretical studies.

As 'higgsteria' swept the world following the CERN announcement. After the announcement of the discovery, the Indian government issued a statement that described Bose as a "forgotten hero."

Bose was born in Calcutta in 1894. He was a lecturer at the Calcutta University College of Science from 1917 to 1921, and then became a reader in physics at the University of Dacca. His greatest recognition came in 1924, when he sent a paper on quantum statistics that had been rejected by a British journal to Albert Einstein who, realizing the importance of the work, translated it into German and published it successfully.

As a result of this recognition, Bose secured a scholarship to travel to Europe, where he conducted research at the Madame Curie Laboratory. He spent time with many other heavyweights in physics, including Einstein in Berlin. Their collaborative work, known as Bose-Einstein statistics, governs the quantum rules for bosons, similar to the way the Fermi-Dirac theorems govern fermions. This work also led to the prediction of a state of matter called the Bose-Einstein condensate formed by atoms cooled to temperatures very near to absolute zero, which was experimentally created in 1995.

It's true that Bose was not directly involved in theorizing the Higgs boson itself. Bose himself was not awarded the Nobel Prize, though it has been given on more than one occasion to the research field initiated by him. Yet, half the particles in the universe obey him. That itself is a remarkable recognition.

Although it is good to acknowledge the contributions of all physicists in the discovery of Higgs boson, science is not a competition, where the frontrunners take pride in winning the medals. Science should transcend all national borders, religions and races. It belongs to all civilizations. The significance of the Higgs boson lies far beyond the superficial discussions that have been rampant in the media, which often obscure the implications of the discovery.

At a deeper level, the real issue at stake is a choice between two different worldviews. The first one posits that matter makes everything, and by the process of reduction, we can reach out to the most elementary particles and their dynamics. Thus particles are the fundamental reality. Our universe and sentient beings are the "effect," caused by the inter-play among these particles – an upward causation.

The other concept relies on downward causation, where consciousness is the funda-mental reality and the ground of all being. This view is close to many spiritual teachings, where free will exists independently and consciousness creates the physical world.

The Hindu Upanishads subscribe to the theory of manifestation and unmanifestation of the universe. The forms and names, including our universe, evolve from an all-pervading cosmic energy. In other words, the material world is a manifestation of the unmanifested. Like waves rise on the ocean, the cosmos comes from the Brahman, the unmanifested.

Although it is a reflection of a deeper thought in tune with Vedantic philosophy that the tangible can spring from the intangible, for all-inclusive Brahman no such distinction between the manifested and the unmanifested exists. Brahman, the eternal and immanent, doesn't distinguish living or non-living, tangible or intangible. Brahman dwells in every-thing from bacteria to humans, in beauty and in the dirt, without being affected by their obvious differences. That's the school of thought of Advaita (non-dualism) Vedanta.

In an attempt to investigate the nature of subtle reality, Vedantic scriptures explain that in the beginning there was One (tad ekam), variously identified as Brahman or Atman. The one divided itself, extending and expanding itself into the vast multiplicity of forms that we can perceive with our senses (manifested), and at the same time the unmani-fested dimension of reality remains beyond perception. The manifested aspect is called vyakta, and the unmanifested facet is avyakta. Both of these dimensions of reality are interconnected.

What Are the Implications of All This?

It would be wild to suggest that the Higgs boson is scientific validation of the story of the universe in Hindu scriptures. However, as we probe deeper into the micro nature of the universe, the feature of interconnectivity becomes more significant, and the separation of the manifested and the unmanifested thins. The Higgs boson is the manifestation of the unmanifested Higgs field and they are interconnected. Vedic thoughts may not define the process of scientific inquiry, but it might, just might, ignite curiosity and higher order thought processes in our journey to understand the universe.

Physicists will be studying the new particle for decades. There could be many varieties of Higgs bosons, or worse, some other unknown particles may be acting as imposters that mimic the Higgs boson.

Even if the newly discovered particle is confirmed beyond the slightest doubt as the previously theorized Higgs boson, there remain an array of questions. As we know, the universe we inhabit is made up mostly of dark matter and is continuously accelerated by dark energy. The Higgs mechanism can explain only the mass of normal matter, which surprisingly is only 5 % of the known universe. Even the standard model with its current framework says nothing about the remaining 95 % of the universe.

Also, while it is comforting to know that the discovery of the Higgs boson that facilitates mass in the universe may now be a known particle, the elementary particles that mediate gravitational force, and are aptly called gravitons, remain unconfirmed. Theoretical physicists generally go beyond the world of particles. For them, the makeup of the universe is something more fundamental than particles. Some think it is strings. Unfortunately, the Large Hadron Collider (LHC) at CERN cannot generate the level of energy required to see such small entities. The LHC is 27 km wide, but we may need colliders the size of the planet Earth itself, a tall order for our current technological standards and abilities, to peek into the deepest secrets of nature.

Besides, describing reality using language and mathematics or any other system of human thinking is imperfect. Are we discovering pieces of reality, or, are we constructing them first and then looking for them? Languages and mathematics may have difficulty in describing the ultimate reality whether it is particles, strings or even information. The history of science and philosophy are punctuated by such fears.

Science describes the world better than any other explanation we have come up with, but we need to recognize the fact that even at its best it cannot explain the universe itself. Our higher order thoughts and mathematics may well enable us to accomplish gene mapping of humans or even the universe, and the Higgs boson discovery nudges us further along in that quest. But, even as we celebrate the discovery, we should also feel a sense of humility.

The human capacity to explore the unknown universe is astounding. Our brains, probably, are even more intricate than the known universe. It's no wonder that we are obsessed with understanding the universe. Here is a remarkable story of a human brain.

THE ODYSSEY OF A BRAIN

It's impossible not to think about the profound mysteries of the universe when we hear the name Albert Einstein. What would you be thinking if you saw pieces of Einstein's brain floating in glass jars?

Two slices of Einstein's brain were on display at Wellcome's new exhibition in London from March 29 to June 17, 2012, named "Mind as Matter." The brain that helped us to demystify space and time continues to attract public attention. The journey of that brain, spanning more than 50 years, is unsurpassed in many ways and is as captivating as its voyage before the death of the genius.

Was Einstein's Brain Preserved After His Death?

Einstein died on April 18, 1955, in Princeton, New Jersey. Thomas Harvey, the doctor who carried out the autopsy on Einstein, put the brain in a jar of formaldehyde and made off with it. Harvey thought he was doing a great service to science by preserving it. He hoped that at some point scientists would be able to figure out the key to the secret of a genius. Later on, the brain was dissected and even sent some pieces to researchers who wanted to study Einstein's brain. However, Harvey kept the jar that contained most of it. The whole brain existed only in photographs taken by Harvey. This entire story is described in the book *Driving Mr. Albert: A Trip across America with Einstein's Brain* (2001) by journalist Michael Paternity.

Is Einstein's Brain So Different?

One paper, entitled "On the Brain of a Scientist: Albert Einstein," appearing in 1985, indicated a greater number of glial cells per neuron in Einstein's brain, which is perhaps linked to his better thinking abilities and conceptual skills. The first anatomical study was published in 1999 by Sandra Witelson, a neurobiologist at McMaster University in Hamilton, Canada. It reported that Einstein's parietal lobes, which are responsible for mathematical thought and visio-spatial cognition, were 15 % wider than normal parietal lobes. Furthermore, other studies have shown that certain parts of the brain were indeed very unusual, though some researchers consider the evidence speculative. So far, the studies haven't established any unusual conclusions to explain Einstein's genius. Though neuroscience has become much more advanced since the time of Einstein's death, it's still a young field, and the "neural basis of intellect" is still not completely comprehended. Further studies might reveal new features of the mastermind involved in imagery and complex thinking.

Where Is Einstein's Brain Now?

Thomas Harvey claimed that he had the authorization to take the brain. Einstein unambiguously directed that his body be cremated after his death, but Harvey claimed he had permission from Einstein's son, Hans Albert Einstein, to conduct a scientific study of Einstein's brain. It's a matter of dispute even now. Harvey moved around the country and always brought the brain with him, although he lost his job and was criticized by many of his colleagues for his actions. In 1998, Harvey returned the remaining parts of Einstein's brain to the pathology department at Princeton University, where the bulk of it remains preserved. Harvey died in 2007. Last year, the Mutter Museum in Philadelphia displayed 46 slides containing slices of Einstein's brain to public view for the first time.

The two slides that are on display in the Wellcome collection are on loan from the Mutter Museum. The brain that embarked on a quest to uncover the unknown universe still remains mysterious. Coincidentally, in Einstein's own words, "The most beautiful thing we can experience is the mysterious. It is the source of all true art and all science."

MIND, MATH, AND MATTER

Strangely, these three words begin with 'm.' But the correlation among them is much deeper than that superficial connection.

Using mathematics as a tool do we discover reality or create new reality? In other words, if you create an abstract mathematical equation, does it truly represent a real thing? These questions clearly divide the mathematicians and both sides have their own arguments.

If the physical world obeys mathematical laws then anything that is made of what makes the physical world should obey the same laws. Why is the brain an exception? A mathematical modeling of brain is possible, and the efforts are on. The working relationship between mind and matter, and expressions to reflect these workings, are possible in the language of mathematics.

The mathematical representations of many worlds and consciousness are not completely available to us now. Although some scientists do not anticipate a mind that can be computed with mathematics, many others are optimistic about such a scenario. Mathematics might hold the key to knowing the thoughts of nature.

> Without mathematics we cannot penetrate deeply into philosophy.
> Without philosophy we cannot penetrate deeply into mathematics.
> Without both we cannot penetrate deeply into anything.
> – Gottfried Wilhelm Leibniz, German mathematician and philosopher

We will continue to debate the origins of mind and matter for long time. Such deliberations are fulfilling in many ways, even if we don't have answers. We can say confidently that the mind-matter problem will occupy a major role in the neurological and psychological sciences of the coming days, and mathematics definitely will be a huge force in bringing them together.

REFERENCES

Ascoli, G. A. (2013). The mind-brain relationship as a mathematical problem. *ISRN Neuroscience, 7*, 1–13, Article ID 261364, 13 pages, 2013. doi:10.1155/2013/261364.

Aurobindo, S. (1972). *The collected works of Sri Aurobindo birth centenary library*. Pondicherry, India: Sri Aurobindo Ashram Publications Department.

Ballentine, L. E. (1970). The statistical interpretation of quantum mechanics. *Reviews of Modern Physics, 42*(4), 358–381. View at Publisher View at Google Scholar.

Barrett, J. A. (1999). *The quantum mechanics of minds and worlds*. Oxford: Oxford University Press.

Bateson, G. (2002). *Mind and nature: A necessary unity (Advances in systems theory, complexity, and the human sciences)*. Cresskill: Hampton Press.

Cleeremans, A. (2005). Computational correlates of consciousness. *Progress in Brain Research, 150,* 81–98.

Douglas, R. J., & Martin, K. A. (2007). Mapping the matrix: The ways of neocortex. *Neuron, 56*(2), 226–238.

Feinberg, T. E. (2012). Neuroontology, neurobiological naturalism, and consciousness: A challenge to scientific reduction and a solution. *Physics of Life Reviews, 9*(1), 13–34.

FQXi Community. (2013). FQXi Community. Available at http://fqxi.org/community/articles/display/178. Accessed 23 Feb 2013.

Hall, M. J. W., & Reginatto, M. (2002). Schrödinger equation from an exact uncertainty principle. *Journal of Physics A, 35*(14), 3289–3303.

Hameroff, S. R., & Penrose, R. (1996). Orchestrated reduction of quantum coherence in brain micro-tubules: A model for consciousness. In S. R. Hameroff, A. W. Kaszniak, & A. C. Scott (Eds.), *Toward a science of consciousness: The first Tucson discussions and debates* (pp. 507–540). Cambridge: MIT Press. Also published in mathematics and computers in simulation (1996) 40, 453–480.

Herculano-Houzel, S. (2009). The human brain in numbers: A linearly scaled-up primate brain. *Frontiers in Human Neuroscience, 3,* 31.

Hesslow, G. (1994). Will neuroscience explain consciousness? *Journal of Theoretical Biology, 171*(1), 29–39.

Jowett, B. (1892). *The dialogues of Plato translated into English with analyses and introductions in five volumes* (3rd ed.). London: Oxford University Press.

King, C. (1991). Fractal and chaotic dynamics in nervous systems. *Progress in Neurobiology, 36*(4), 279–308.

Max Planck Florida Institute – About Us: About Max Planck Florida Institute. (2013). *About Max Planck Florida Institute.* Available at http://www.maxplanckflorida.org/about.html.

Paterniti, M. (2001). *Driving Mr. Albert: A trip across America with Einstein's brain.* New York: Dial Press.

Penrose, R. (1994). *Shadows of the mind; an approach to the missing science of consciousness.* Oxford: Oxford University Press.

Penrose, R. (2002). *The Emperor's new mind: Concerning computers, minds, and the laws of physics (popular science)* (1st ed.). New York: Oxford University Press.

Penrose, R. (2005). The road to reality: A complete guide to the laws of the universe. In A. A. Knopf (Ed.), *Sketches (Canto classics)* (Reprintth ed.). New York: Cambridge University Press.

Penrose, R. (2011). Consciousness in the universe: Neuroscience, quantum space-time geometry and Orch OR theory. *Journal of Cosmology, 14.* Available at http://journalofcosmology.com/Consciousness160.html. Accessed 11 May 2012.

Poe, E. A. (1845). The Raven. *American Review, 1,* 143–145.

Putnam, H. (1979). 'What is Mathematical Truth' and 'Philosophy of Logic'. In: *Mathematics matter and method: Philosophical papers* (2nd ed., Vol. 1). Cambridge: Cambridge University Press.

Schrodinger, E. (2012). *What is life?: With mind and matter and autobiographical Sketches (Canto classics)* (Reprint ed.). Cambridge: Cambridge University Press.

Sporns, O. (2010). *Networks of the brain* (1st ed.). Cambridge/Massachusetts: The MIT Press.

Tononi, G., & Edelman, G. M. (1998). Consciousness and complexity. *Science, 282*(5395), 1846–1851.

Weerasinghe, S. G. (1993). *The Sankhya philosophy: A critical evaluation of its origins and development.* New Delhi: South Asia Books.

Wigner, E. P. (1967). *Symmetries and reflections.* Cambridge: MIT Press.

Witelson, S. F., Kigar, D. L., & Harvey, T. (1999). The exceptional brain of Albert Einstein. *The Lancet, 353*(9170), 2149–2153.

10

The Spooky World of Quantum Entanglement

> *"Anyone who is not shocked by the quantum theory has not understood it."*
>
> – Niels Bohr, Danish physicist who received Nobel Prize in 1922 for his work on the structure of atoms.

Why can't we be in two places simultaneously? Why can't we communicate instantaneously? Better yet, why can't we teleport ourselves to another location instantly?

BREAKING THE RULES

Actually, subatomic particles can. Unfortunately, even though we are comprised of these very particles, big objects don't follow the rules of quantum mechanics, which, at the subatomic level, make possible such incredible feats. But, the weird nature of quantum behavior explicit in the subatomic world is gradually encroaching our classical world. Researchers are in pursuit of applications that would assist us in overcoming space-time constraints and the laws of classical physics. Future generations might see teleportation and a universal quantum web as a normal aspect of their daily lives.

The underlying phenomenon that enables the seemingly impossible is quantum entanglement – the mysterious way by which two particles are connected instantaneously irrespective of their spatial distance. Entanglement occurs when two particles (photons, atoms, electrons, etc.) become so deeply linked together that the changes that happen to one are instantly reflected by the other even if they are residing in opposite corners of the universe. This means that any information imposed on one particle can be easily retrieved from the other without physically sending it across the distance that divides them. Not surprisingly, researchers and governments are frenetically chasing practical applications of entanglement, such as entanglement-assisted cryptography and teleportation.

S. Mathew, *Essays on the Frontiers of Modern Astrophysics and Cosmology*, Springer Praxis Books, 149
DOI 10.1007/978-3-319-01887-4_10, © Springer International Publishing Switzerland 2014

Figure 10.1. Quantum delocalization. Quantum mechanics tells us that lighter elements, such as hydrogen, are delocalized in space (CC Image courtesy of UCL Mathematical and Physical Sciences on Flickr. Used with permission).

In 2012, researchers at the University of Vienna in Austria set the record for transmitting entangled photons over a distance of 90 miles (Song Ma et al. 2012). The distance achieved in the experiment illustrates the possibility of quantum communication in space involving satellites. Their Space-QUEST (Quantum Entanglement for Space Experiments) proposal offers a step in that direction.

On Earth, which is less than 8,000 miles in diameter, we think our communications are instantaneous, as signals travel at 186,000 miles per second, the speed of light. But consider the New Horizons spacecraft, which was launched in 2006 toward the boundary of our Solar System to study Pluto. This spacecraft, cruising at thousands of miles per hour, will reach Pluto in 2015. At a distance of more than 3 billion miles, its communication with Earth would take a few hours. If that is the timescale in our own backyard in the universe, imagine future missions to farther destinations. Perhaps future civilizations will master the techniques of entanglement to communicate and transport instantaneously.

Entanglement defies some of our deepest perceptions about the nature of reality. It does not place a cap on speed, which Einstein's special theory of relativity constrained to the speed of light. Instead, one particle (location) mysteriously shares the information with its entangled partner (another location) instantaneously. Information is not carried by electromagnetic waves at the speed of light, which is finite, although it is very high. Instead, by observing one particle, we instantly get the information about the other particle even if it is light-years away. Until such an observation (measurement) is made, no information exists – or, in the weird realm of quantum physics, all possible information exists.

Long accepted in microscopic systems, quantum entanglement has more recently been observed at the macroscopic level as well. In one recent study, John Martinis and his colleagues at the University of California used two superconductors to send out streams of entangled electric currents. According to the laws of classical physics, the directions of the currents should have been completely independent of each other. But each time Martinis and his team measured the direction of the current, they observed that when one current flowed clockwise, the other flowed counterclockwise, demonstrating that both streams are linked to each other inexplicably. It was one of the most visible illustrations of quantum entanglement in the classical world.

We presently have no explanation for such entanglement. In the orthodox version of quantum physics, objects do not exist or have any inherent properties, and their existence becomes real only when minds interact with them through observation or measurement. Consequently, reality is created by interaction. Admittedly, until that interaction happens, all possibilities exist as the superposition[1] of wave functions.[2] All the available states that are represented by various wave functions collapse to create a single measurable form of reality. In other words, when the interaction happens all the possible states disappear except the one that is observed, and we interpret that as the reality.

One of the leading figures of quantum physics, David Bohm, had a more deterministic approach, arguing that the quantum world can exist independently of the human mind (Weber 1986): "In some sense, man is a microcosm of the universe; therefore what man is, is a clue to the universe. We are enfolded in the universe. Indeed, the attempt to live according to the notion that the fragments are really separate is, in essence, what has led to the growing series of extremely urgent crises that are confronting us today."

Many scientists believe that some hidden variables are behind the phenomena of entanglement. Einstein, who sought the ultimate truth through mathematical equations, ridiculed the concept of "the spooky action at a distance," dismissively remarking, "I like to think that the moon is there even if I am not looking at it." Yet, even he admitted to the evidence of entanglement, whose existence has been established beyond any doubt through numerous experiments.

According to cosmologists, the universe began as a quantum fluctuation in a limitless void. The universe was a huge quantum superposition of all possible states until the first primordial mind observed it, causing it to collapse into one reality, eliminating all other probable states. In one version of quantum mechanics, each action opens up another universe, thereby creating an infinite number of parallel universes. Some physicists extend this argument even further, saying that the creation of our universe from a singularity demands that everything that was created later (stars, planets, plants, animals, and humans) became entangled with each other; the super entanglement that existed in the beginning was not broken but only spread to a larger extent. This explanation is very encouraging for proponents of paranormal phenomena, such as telepathy, which most scientists dismiss, who claim that the cells of our brain are entangled with everything else in the universe, thereby making it possible to decode information received from any other object or part of the universe.

---

[1] The property of a particle or object to exist in all the theoretically available states.

[2] In quantum mechanics wave function is the mathematical description of a particle or an object.

Whatever the explanation for the quantum entanglement phenomenon, it has exciting real-life applications. One immediate area where entanglement can play a crucial role is in computing. Computer binary codes of "1" and "0" are used in information exchanges in the digital age, which provides opportunities for eavesdropping or hacking of information. However, quantum entanglement provides a better encryption tool using the properties of photons, such as spin. The traditional "bit" is replaced by "qubit" (quantum bits) to represent quantum information. If an attempt is made to intercept a photon-based message, the spin changes, signaling a possible compromise in security. MagiQ Technologies, Inc., of Cambridge, Massachusetts, sells technology built on quantum cryptography to banks and corporations. The encryption refreshes its quantum keys as often as 100 times a second during a transmission, making it an extremely hard encryption to break.

However, even though quantum encryption is possible, presently quantum computing is constrained by many practical challenges, including the breakdown of the spooky relation in systems because of interaction with the external world. So widespread quantum computing is a long way off. Theoretically speaking, quantum computers would be far superior to classical ones. Since they can hold all available values in qubits, rather than classical bits, they can search the right information through huge databases faster and much more efficiently.

If quantum entanglement's promise for information exchange is exciting, the possibility of quantum teleportation is downright dizzying. As the famous science writer Arthur C. Clarke once observed (1984): "Any sufficiently advanced technology is indistinguishable from magic."

In quantum teleportation, properties such as the spin of a particle or the polarization of a photon are transferred from one place to another without traveling through a physical medium, making it instantaneous. In 1993, an international group of scientists confirmed the practical possibility of quantum teleportation – an idea that until then had been exploited mostly by science fiction writers (Bennett et al. 1993). Since then, entanglement-assisted teleportation has been tried successfully by different groups of researchers. Much as a fax machine scans and sends information to another location, hypothetical teleportation machines would reproduce an exact copy rather than an approximate duplicate.

In quantum teleportation, no material is sent from one location to another. Instead, quantum entanglement is used to impose information onto a particle at one location and is retrieved through a different particle, which is entangled with the first particle, at another location. When all the information about the first particle is obtained at the other location, what is transported is the first particle itself. Strangely enough, the original is destroyed in the process of quantum teleportation. In 2009, a team of scientists (Olmschenk et al. 2009) from the Joint Quantum Institute (JQI) at the University of Maryland and the University of Michigan succeeded in teleporting a quantum state directly from one atom to another over a substantial distance, getting a step closer to practical quantum teleportation. The group reported that in their protocol, atom-to-atom teleported information can be recovered with perfect accuracy about 90 % of the time – and that figure can be improved.

Definitely, technological and engineering challenges have to be met before these experimental results become any feasible mechanism. However, that is the case with many of the inventions in the past. For example, picture the first crude solid state transistors and then think about how they shaped our lives in later years.

Researchers hope to achieve the teleportation of complex systems, such as a virus, in coming decades. Applying the process to larger objects, including people, is largely a scientific engineering problem, which "is likely to be solved in time," according to Michio Kaku, author of the bestseller *Physics of the Impossible*. Teleporting humans would involve scanning all the information in trillions and trillions of atoms that make up the person, but the technology to perform such a task does not exist at this time and may not be available for centuries to come.

Scientists cannot explain why entanglement is not visible outside their labs right now. But future generations might communicate instantly across the universe and may appear and disappear at will, no doubt a magical phenomenon by today's standards. When we do master quantum entanglement, we may well be led to the eternal truth advocated in ancient philosophies and religions – that we live in the cosmos and the cosmos lives in us, that the space and time that separate us are just creations of the mind. As the Buddhist *Mahayana Sukta* says: "All such notions as causation, succession, atoms, primary elements that make up personality, personal soul, Supreme Spirit, Sovereign God, Creator, are all figments of the imagination and manifestations of mind" (Conners 2011).

THE NATURE OF REALITY

The modern scientific interpretation that our observation creates physical reality seems rather absurd. One can dismiss that seemingly strange argument with the famous Eiensteinian answer that the Moon is still there even if we don't look at it. However, as a paradigm shift sweeps through the world of science, much to our dismay, we may have to disagree with the notion of the iconic scientist. An increasing number of modern physicists are jumping on the bandwagon piloted by the quantum view, once considered to be an offshoot of mainstream physics, and they are ready to wrestle with the implications of that bizarre aspect of modern physics. And, if they are right, we, the sentient beings, are shaping our future as well as our past by contributing to this participatory universe.

The mysterious relationship between observation and its outcome dates back to the early days of quantum mechanics. Austrian physicist Erwin Schrödinger demonstrated this relationship with a thought experiment popularly known as the Schrödinger cat. He imagined a cat in a closed chamber along with a vial of poisonous gas. The chamber also contains a radioactive element that may emit a particle. If the particle is emitted, a triggering mechanism would release the poisonous gas that will kill the cat. If not, the cat will stayed alive. In order to emphasize the conflict between quantum and classical worlds, Schrödinger posed the question, "What could be known about the cat before opening the box?".

The simplest and most logical answer is that the cat may be dead *or* alive. But, according to quantum mechanics, the cat is *both* dead and alive. We know that neither answer is true and real until someone opens the box and looks inside it. This quantum paradox, like many others, demonstrates the limitations of our knowledge when dealing with the outside world. Until a definite conclusion is made, we have to cope with all the possibilities – in other words, uncertainty rules. Of course, Schrödinger was trying to prove inconsistency when the quantum view is extended to the classical world.

Figure 10.2. Schrödinger's cat. The cat is both dead and alive because of the separation of the universe due to two superimposed and entangled quantum mechanical states. When the observation is made the wave function collapses and the cat is either dead or alive.

However, the important question is, are we influencing the outcome with our observations? This has been proved beyond doubt in the experiments with fundamental particles such as electrons. Quantum mechanics has been credited as the most valuable creation of the scientific adventure. Originally designed to account for the behavior of fundamental particles, it has shown incredible strength in its predictions and endured every experiment designed to prove or challenge its calculations. But many of its predictions, such as the cat experiment, defy common sense – a reason why Einstein distanced himself from the proponents of quantum physics though he played a major part in its creation. He believed in a universe that has an underlying order and beauty, one that could be explored with mathematical equations.

Einstein summarized his dislike of the uncertainty and unpredictability of quantum mechanics by saying, "God does not play at dice with the universe." But Neils Bohr responded, "Quit telling God what to do!" Quantum physicists were adamant in their view of the uncertain world. They seem to follow what Krishna says in the Bhagavad Gita: "I am the game of dice. I am the self-centered in the heart of all beings."

The heart of quantum mechanics is a wave function, a mathematical description of the physical system. In the case of the Schrodinger cat scenario, a wave function represents every possible state of the inhabitants of the chamber. The wave function exists as a superposition of all the available possibilities, meaning the cat could be dead or alive or it could exist as a combination of every state between dead and alive. The observer will never be able to see simultaneously all the possibilities. When the observer does see the system, the collapse of the wave function occurs, resulting in only a single outcome – dead or alive.

If one could extend that argument, it leads to the question of whether the universe exists if we don't look at it. When physicists say there must be an observer to look at the universe for the universe to exist, does that demand conscious observers for the existence of the universe? Not necessarily, since matter and radiation could play the same role as a conscious observer; the universe is a big quantum system. What we see is just one of the possible outcomes that happened to be our 14-billion-year-old universe. As observers, we are ignorant about the other possibilities and universes with different histories and beings. Our observation collapsed the other states that existed along with this present one.

The light we see from distant galaxies does not create a reality unless it has interacted with matter, whether it is with the atoms of our own eyes or with machines such as telescopes. This is what many physicists consider observation, not the mere act of looking at it by conscious beings. In the absence of interaction between energy and matter, there is no universe or even planets or beings.

John Wheeler, the late pontiff of quantum mechanics, explained it in this way. According to him, most of the universe consists of huge clouds of uncertainty that have not yet interacted either with a conscious observer or even with some lump of inanimate matter. He sees the universe as a vast arena containing realms where the past is not yet fixed. When we look at it, what manifests is a single reality among the infinitely possible realities. The possible outcomes of countless interactions become real, where the infinite variety inherent in quantum mechanics manifests as a physical cosmos. And we see only a tiny portion of that cosmos.

The objective reality that we thought to exist is blurred into the subjective consciousness in the territory of quantum mechanics. We have always been taught that there exists an absolute reality independent of our actions. It is often paired with unlimited beauty, ultimate truth, and absolute knowledge. Accomplishing that pure reality, according to Hindu philosophy, would bring nirvana – the ultimate liberation and subsequent union with the Supreme Being.

Absolute truth and beauty is appealing to the human psyche. Art, religion, and other human endeavors emphasize the place of an entity linked to the supreme qualities in every culture. Some call it the impersonified God, as Gods are the manifestations of that absoluteness. The modern view of science, as interpreted by quantum physics, suggests that even that absoluteness is our own creation.

It's been said that there are only two possibilities about the nature of reality. Either it is impossible to ever know about it, or a theory will describe it. Science has long accepted this ambiguity as it experienced the enormity of the mysteries presented by the cosmos. While it continues its efforts to comprehend that enormity, it also creates vistas of reality where we are the masters of creation. The newly emerging picture of the cosmos portrays beings as not just passive observers of the universe but the ones who participate in the creation of the universe and are also part of the universe.

Science, through its rational and logical process, is seeking an ultimate theory, a theory of everything, that would explain everything in the universe including our own existence and one that could answer why the universe exists. The same science that believed in an absolute reality-based model of the universe is now gearing up to shed that notion, heralding a new era in science. In this era, there is no single reality, but rather it is our obseravtion that creates reality. That would mean there is no single truth or a single reality independent of our observation and the subsequent models we create.

Are such thoughts and ideas planting the seeds of destruction of the elegant model of reality we always sought? Is the harmony and beauty that Einstein thought of as the underlying principle of the universe at risk?

In quantum mechanics, objects can be in more than one place at the same time. A particle can penetrate a barrier without breaking it. Something can be both wave and particle at the same time. In fact, observation distorts the reality and creates a new one.

Almost everything that involves atoms and molecules depends on quantum mechanics, from transistors to lasers. It's been said that one-third of the U. S. economy depends on devices based on quantum mechanics. Quantum mechanics also tells us weird things about our world – and maybe about us and our perceived reality and consciousness.

Again, Oxford physicist Roger Penrose argues that there is a world beyond the material one, and it is linked to our consciousness. The quantum or classical world can't completely explain consciousness. The physics that we understand at the moment is simply not rich enough to incorporate the phenomenon of consciousness.

It looks like we are forced to accept the doctrine that the universe is not made of material bits but with the bits of thoughts.

According to ancient Hindu philosophy, the whole universe is made up of five basic elements: earth, water, fire, air, and space. Every living or non-living thing is also made up of these five basic elements of nature. These are referred to as the pancha-mahabhootas and are related to our senses as well. As science pursued and purified the constituents of the universe, we learned that atoms and fundamental particles make up our universe and its components. Later on, the particles, at least theoretically, give way to strings of energy as the constituents of our cosmos. Modern interpretations of quantum theory go beyond even these traditional explanations. Some researchers suggest that the physical universe is made of information, not particles or strings. As echoed in Wheeler's comment, "it from bit," where *it* refers to the material world and *bit* is the information.

Will these questions – From what came existence? What is the nature of reality? – ever be answered? Physicists are skeptical. That skepticism has a valid rationale; why do we have to be arrogant in our quest for the truth? As a species, we are relatively new, and our scientific understanding began to unfold just over a few 100 years ago. We are not at the point where we can demand to understand everything.

We are, definitely, not the last link in evolutionary intelligence, so it is arrogant on our part to insist on the final answer. There are some domains of the universe beyond the tools of our current science. All we can do is continue to explore and celebrate the vast possibilities presented to us with the hope that 1 day we can understand everything.

Quantum mechanics and its theoretical connotations are not the only strange phenomena in physics. The dance of fundamental particles that make up our reality is an exciting topic to study. In recent times, the Large Hadron Collider (CERN) and the experiments conducted there have become very interesting to the public.

LARGE HADRON COLLIDER BEGINS WITH THE BEGINNING

No other experiment in human history has captured the imagination of scientists and common folks alike the way the Large Hadron Collider (LHC) has done.

The gigantic collider underneath the Alps, just outside of Geneva, has finally accomplished the landmark collision of proton beams. The LHC successfully smashed the proton beams at 7 Tev (Tera electron volt). Scientists described the event as "a new era in science." Scientists hope that the collisions will allow them to answer questions such as those surrounding super symmetry, hidden dimensions, and the existence of the Higgs boson.

The standard model of particle theory postulates that everything is made up of fundamental particles called fermions (like electrons, the basic unit of charge) and bosons (like photons, the basic unit of light). No one disagrees with this theory. In fact, we manipulate these particles in such a way that they have been employed to control anything from electronic circuits to solar cells.

However, there is a missing particle in the standard model – the Higgs boson – the "God particle." The theory claims that Higgs bosons are responsible for providing the mass to everything in the universe. Some particles are massive while others have negligible mass, but the mass in Higgs bosons has eluded us for a long time.

High-energy collisions can create Higgs bosons, though they last only for a split second before decaying into other particles. Nevertheless, they can be traced back from the trails of their remnants. The detection of the Higgs boson can help explain the origin of mass in the universe, an incomplete chapter in the field of particle physics thus far.

Haunted by mechanical problems and doomsday prophecies, LHC has struggled since its first major operation in 2008. Apparently, some scientists even suggested that "the Higgs boson doesn't want to be found" by feeble humans, that the God particle would actually be aware enough to go back in time and prevent its own creation!

Holger Bech Nielsen, a Danish theoretical physicist, and Masao Ninomiya, a Japanese physicist, suggested that the Higgs boson, the particle that physicists think they have created with the collider, might be "abhorrent to nature." They claimed that influence from the future would bring bad luck for the LHC. It seems that even professional scientists are not free from spooky speculations (Harrell 2009).

Researchers are relying on LHC for another reason as well – super symmetry (SUSY). The detectors in the experiment will look for super-symmetric particles. Super symmetry hypothesizes that for every standard model particle, there is a corresponding super-symmetric particle. Thus the electron will have a super-symmetric partner called a selectron, with a different spin.

If these "sparticles" exist as predicted by theory, the LHC will provide us with an opportunity to see them for the first time. While the ordinary particles will be delighted to see their super partners, researchers will attempt to answer the dark-matter mystery.

Super-symmetric particles are the leading candidates for demystifying dark matter, the exotic matter that makes up roughly 23 % of the universe. The ordinary matter that creates the galaxies, stars, planets, and different forms of life is only a tiny fraction of the universe, estimated to be about 4 %. The presence of dark matter is inferred only through its gravitational effect, as it doesn't emit any electromagnetic radiation, like ordinary matter does.

The first attempt to link the LHC with disaster was that it might trigger a black hole that would swallow our planet. That fear even sparked a lawsuit in a U. S. court, only to be dismissed later. Scientists say that fears such as these are unjustified. In fact, cosmic ray collisions that take place in the upper atmosphere of our planet have been known to occur at high energies and have been well tolerated by our planet for the last 4.5 billion years.

Now, if a micro black hole is produced, which is highly unlikely, it would be a fascinating opportunity for researchers to study it. These micro black holes will decay quickly without any harm. But the alarmists are not ready to accept that verdict.

Perhaps the most exotic discoveries from this experiment are the extra dimensions. Our world of three spatial dimensions and one dimension of time is not adequate enough to accommodate many observed effects. String theorists are betting that the universe must unfold around ten or eleven dimensions. Unfortunately, they are unable to provide any clarity about the nature of these dimensions.

Attaining energy levels that no other collision ever achieved has the potential to open doors to other dimensions. This may explain why gravity is weak in our universe as compared to other forces; gravity might be operating in other extra dimensions. Interestingly, scientists have learned to live with extra dimensions and parallel universes. As such, we might be the inhabitants of a 3D blob, one of the innumerable multiverses.

The LHC will run for many months, giving the detectors ample data to analyze. This will take months or even years with all of the latest computing techniques. Until then, no one will know the new vistas generated in the particle kingdom. After this period, the LHC will shut down for up to a year to gear up for further collisions with more energy.

The collisions of the particles in the LHC are truly the minuscule replication of the events that unfolded in the early universe about 13.7 billion years ago, immediately following the Big Bang. The LHC is recreating those events in a unique way. It is unique because every story has a beginning, but the LHC is trying to tell us the story of the beginning. Nonetheless, this story has many twists and turns, including the violation of a fundamental law. So we could say that we are the children of violation.

THE CHILDREN OF VIOLATION

In our everyday life, violations of enforced law result in tickets and can often advance to more intricate situations. However, physicists point to the violation of a physical law to billions of years ago that eventually led the creation of everything, including us. We are essentially the children of that violation.

Physicists believe that at some point in the early moments of our universe, a symmetry violation occurred, paving the way for matter to win out over antimatter particles. As time progressed, this triumphant matter manifested in different forms such as stars, planets, life, and people. In that sense, we are obliged to this violation that prompted even our own existence.

The latest results from experiments provide clues to understanding this violation. Recently, scientists at the Relativistic Heavy Ion Collider (RHIC) reported the findings of their research carried out in the 2.4-mile particle collider located in Upton, New York. This "atom smasher" recreates the conditions similar to those that existed in the early universe. The data from the RHIC experiments fascinatingly point to a possible parity violation in strong interactions that bind the quarks and gluons.

Subatomic particles such as neutrons and protons are actually complex in nature. Both of these nucleons are composed of three quarks and gluons that bind the quarks together.

At the RHIC experiment, quarks and gluons are ripped from their parent nuclei to create a hot soup of quark-gluon plasma. Physicists believe that matter existed in a physical state identical to this in the universe micro-seconds after the Big Bang.

Researchers recreate this condition at RHIC by using high-energy collisions of the nuclei of gold atoms. The resulting temperature of their latest experiment measured 4 trillion degrees Celsius – the hottest temperature ever attained in the laboratory. The liberated quarks and gluons behaved more like a liquid, with considerable cooperation among them. This "new material" and its properties might be hiding the secrets of our cosmic origin.

The standard model of particle physics demands a symmetry known as charge-parity (CP) to be conserved in the universe. The laws of physics must remain invariant for particles under charge exchange (C symmetry) or left and right swapping (P symmetry). This implies that our world is indistinguishable from its mirror image. However, parity violation was identified in the 1950s in weak interactions such as beta decay.

The strong interaction that operates between quarks and gluons opposes parity violations under ordinary conditions. The data from the RHIC experiments intriguingly points to a possible parity violation in strong interactions. This cannot be perceived as breaking the known laws of physics, since these very same laws predicted such effects. Yet, it may be helpful to learn more about the earliest violation that shaped our cosmic history.

According to our current understanding, there existed an equal number of particles and antiparticles in the early universe. Equal numbers of matter and antimatter particles completely annihilate each other to produce energy and conserve symmetry. Yet, we know that in our universe, matter dominates antimatter.

If this symmetry had never been broken, matter and antimatter would have completely annihilated each other, leaving the whole universe as pure energy. Though scientists will not be able to explain the global dominance of matter over antimatter with the information from their latest results, it will definitely assist them in searching for the answer.

In fact, physicists celebrate when the laws of nature are violated. For when this happens, it provides physicists an opportunity to search for a deeper understanding of nature. As a result, new concepts and theories of physics can emerge.

The increased understanding attained concerning the quark-gluon interactions in the RHIC experiments will assist scientists in probing deeper, and they can transfer the gained knowledge to LHC experiments at CERN.

Once the imprints of broken symmetry become evident, we can reconstruct the puzzles of our early cosmic history. That would further underscore the significance of symmetry violation in the cosmic creation.

People often ask, "Why do we pump billions of dollars into these high energy experiments? Who cares about quarks and gluons?" The usual answer is, "We're trying to answer the fundamental questions," which sounds like a cliché. That may be pleasing to scientists but not to taxpayers.

The truth is that the technological spinoffs from these experiments are present in our everyday life; from superconducting materials used in Maglev trains to PET scans in our hospitals, all bear a common thread. Their origins can be traced back to these experiments. Even the modern day Internet was born of out necessity in particle experiments.

THE MAN BEHIND THE CAT: SCHRODINGER

In an earlier section, we mentioned the metaphorical Schrödinger cat. Perhaps the cat is more popular than Schrödinger himself! So let's look briefly at the life of the man who created the mystical cat. Erwin Schrödinger was born on August 12, 1887, in Vienna, Austria. He received the Nobel Prize together with Paul Dirac in 1933 for the "discovery of new productive forms of atomic theory."

Schrödinger's contributions were not just limited to quantum mechanics. He made contributions to different branches of science and had a deep interest in philosophy. His book *What Is Life* was an attempt to explain biological phenomena based on laws of physics such as the second law of thermodynamics.

In our previous discussions, we have seen debates between physicists about the nature of reality. Einstein and Schrödinger were critical of the randomness and probability implied by quantum mechanics to describe nature. They held a deterministic view regarding the nature of the universe. The famous Schrödinger's wave equation predicts a perfectly deterministic time evolution of the wave function as opposed to the probabilistic view advocated by Neils Bohr and Werner Heisenberg.

Figure 10.3. A man can be buried but not his equations – Mortal man and immortal equation. The gravesite of Erwin Schrödinger's and his wife Annemarie; above the name plate Schrödinger's quantum mechanical wave function is inscribed (Image courtesy of Rodney Jones, 2005, Randburg, South Africa).

Each particle can be represented by a wave function, which predicts the probability of an outcome. What does Schrödinger's equation tell us about physical reality?

Unfortunately we can't derive definitive information from the solutions of this equation, like in classical physics. The equation only provides the probability of finding the particle in a given location at a time. Schrodinger's equation describes how a physical system changes over time. Indeed, it is considered as one of greatest achievements of physics. It describes how a physical system will change over time.

We already discussed the dual nature of particles, which means that particles can exhibit behavior normally attributed to waves. Since wave function considers particles as waves it favors the idea that wave nature is closer to physical reality. The properties of particles manifest under certain conditions.

In fact, the Schrodinger cat scenario was introduced to point out an inherent paradox that lies at the heart of the Bohr-Heisenberg doctrine. Along with Einstein Schrodinger did not support the existence of all possible states that was popularized by the Bohr-Heisenberg duo.

Schrodinger died on January 4, 1961 (aged 73) in Vienna, Austria.

REFERENCES

Barrow, J. D., & Tipler, F. J. (1988). *Quantum mechanics and the anthropic principle. The anthropic cosmological principle.* New York: Oxford University Press.

Bennett, C. H., Brassard, G., Crépeau, C., Jozsa, R., Peres, A., & Wootters, W. K. (1993). Teleporting an unknown quantum state via dual classical and Einstein-Podolsky-Rosen channels. *Physical Review Letters, 70,* 1895–1899.

Bohm, D. J. (1975). *On the intuitive understanding of nonlocality as implied by quantum theory* (pp. 96–102). Netherlands: Springer.

Clarke, A. C. (1984). *Profiles of the future: An inquiry into the limits of the possible* (Rev Sub ed.). New York: Henry Holt & Co.

Cohen-Tannoudji, C.(1992). Quantum mechanics (2 vol., Set ed.). New York: Wiley-VCH.

Conners, S. (2011). Zen Buddhism – The path to enlightenment – Special Edition: Buddhist Verses, Sutras & Teachings. El Paso/Texas: Special Edition Books.

Conze, E. (1958). *Buddhist wisdom books, containing the diamond Sutra and the heart Sutra.* London: G. Allen and Unwin.

Does the universe exist if we're not looking?. Does the universe exist if we're not looking?. http://discovermagazine.com/2002/jun/featuniverse/article_view?b_start:int=2&-C=. Accessed 29 Oct 2009.

Everett, H. (1957). 'Relative state' formulation of quantum mechanics. *Reviews of Modern Physics, 29,* 454–462.

Gleick, J. (2008). *Chaos: Making a new science* (Revised ed.). New York: Penguin Books.

Gribbin, J. (1984). *In search of Schrödinger's cat: Quantum physics and reality.* Toronto/New York: Bantam Books.

Gribbin, J.(1996). Schrodinger's kittens and the search for reality: Solving the quantum mysteries. New York: Back Bay Books.

Gribbin, J. (2010). *In search of the multiverse: Parallel worlds, hidden dimensions, and the ultimate quest for the frontiers of reality* (1st ed.). Hoboken: Wiley.

Gribbin, J. (2013). *Erwin Schrodinger and the quantum revolution* (1st ed.). Hoboken: Wiley.

Harrell, E. (2009). *Large Hadron Collider: Damaged by a time-traveling bird?* – Time. Available at http://www.time.com/time/health/article/0,8599,1937370,00.html. Accessed 10 Dec 2009.

Landau, L. D., & Lifshitz, L. M. (1981). Quantum mechanics non-relativistic theory (3rd ed., vol. 3). Burlington/Massachusetts: Butterworth-Heinemann.

Moore, W. (1989). *Schrödinger – Life and thought*. Cambridge: Cambridge University Press.

Olmschenk, S. et al. (2009). Quantum teleportation between distant matter qubits. *Science, 323*(5913), 486–489.

Quantum Fluctuation. A Review of the universe. http://universe-review.ca/R03-01-quantumflu.htm. Accessed 4 Feb 2011.

Richard, L. (2002). *Introductory quantum mechanics* (4th ed.). Reading: Addison-Wesley.

Song Ma, X. et al. (2012). Quantum teleportation over 143 kilometres using active feed-forward. *Nature, 489*(7415), 269–273. Available at http://www.nature.com/nature/journal/v489/n7415/full/nature11472.html. Accessed 19 Oct 2012.

Weber, R. (1986). Dialogues with scientists and sages: Search for unity in science and mysticism. Law Book Co of Australasia.

Williams, C. (Ed.). (1999). *Quantum computing and quantum communications, vol. 1509, Lecture notes in computer science*. Berlin: Springer-Verlag.

Williams, P. (2005). The origins and nature of Mahayana Buddhism; some Mahayana religious topics, (1. published. ed.). London/New York: Routledge.

Zajonc, A., & Houshmand, Z. (2004). *The new physics and cosmology: Dialogues with the Dalai Lama*. New York: Oxford University Press.

Zukav, G. (1979). The dance. In *The dancing wu li masters* (pp. 237–279). New York: Perennial.

11

Are We Alone?

> *FM signals and those of broadcast television...[travel] out to space at the speed of light. Any eavesdropping alien civilization will know all about our TV programs (probably a bad thing), will hear all our FM music (probably a good thing), and know nothing of the politics of AM talk-show hosts (probably a safe thing).*
>
> – Neil De Grasse Tyson in his book
> *Death by Black Hole* (2007)

The question "Are we alone?" has been lingering for a long time, probably since people began gazing at the sky.

FERMI'S PARADOX

We will begin this discussion with a lunchtime conversation that happened in the summer of 1950 at Los Alamos. A group of veteran physicists had assembled there to provide key instrumentation for the upcoming thermonuclear test of the hydrogen bomb. The eminent physicist Enrico Fermi had a penchant for posing interesting questions during the lunches.

At one point, Fermi and his colleagues had a casual discussion on life, intelligence, and alien life. Referring to the existence of aliens Fermi asked, "Where is everybody?" He was mentioning the apparent contradiction between the speed of colonization, if advanced aliens exist, and the lack of evidence of extraterrestrial beings. Far from being the beginning of a joke, this question and discussion have come to be known as Fermi's Paradox. If the universe is as old as it is, and if Earth isn't the oldest planet with intelligent life, and conquering the galaxy is as easy as it seems, then where the heck are they? Fermi's opinion was that we are alone. So far Fermi is right, we should say.

S. Mathew, *Essays on the Frontiers of Modern Astrophysics and Cosmology*, Springer Praxis Books, DOI 10.1007/978-3-319-01887-4_11, © Springer International Publishing Switzerland 2014

Since the time of Fermi, many solutions have been proposed to resolve the Fermi paradox. Nevertheless, the actual action began about a decade later. A young American astronomer named Frank Drake steered a big radio telescope to a nearby star system, eager to listen to an alien world. He didn't heard any voices at that time and, so far, even after five decades, the answer is still an uncanny silence.

Frank Drake's mission to embark upon the alien hunt has since grown into a large enterprise, beginning its operations in 1985. Known as SETI (Search for Extraterrestrial Intelligence), the core premise of the search is based on the assumption that our cosmos is teeming with life – not in its primitive form, but technologically sophisticated forms, similar or possibly more advanced than us.

The Drake equation estimates the number of detectable civilizations in the observable universe as something around 10,000, with the nearest one at least about 1,000 light-years away. That alone speaks to the major difficulty in the alien search – literally distances of astronomical proportions.

The Drake equation is usually written as:

$$N = R^* \cdot fp \cdot ne \cdot fl \cdot fi \cdot fc \cdot L$$

where

N = The number of civilizations in the Milky Way Galaxy whose electromagnetic emissions are detectable.

R^* = the average rate of formation of stars suitable for the development of intelligent life.

fp = the fraction of those stars with planetary systems.

ne = the number of planets, per solar system, with an environment suitable for life.

fl = the fraction of suitable planets on which life actually appears.

fi = the fraction of life bearing planets on which intelligent life emerges.

fc = the fraction of civilizations that develop a technology that releases detectable signs of their existence into space.

L = the length of time such civilizations release detectable signals into space.

As you can imagine, the parameters in the equation vary widely and so does the result. The estimates put the number of civilizations from few hundred to thousands depending on the value of the parameters. Nevertheless, the implications of finding other civilizations would be fascinating!

The Drake equation with so many uncertain factors can yield a number of different answers with some in the range of 10,000 (Nadis 2010). This means about 10,000 communicative civilizations possible in our own galaxy.

The importance of this equation is not to calculate an absolute number, but it opens up a whole new opportunity to explore the universe. This could be a great opportunity to contemplate about other beings that are beyond the constraints of space and time.

It is assumed that advanced civilizations are capable of using electromagnetic waves to communicate, or even to get our attention. Consequently, researchers are looking for "narrow band signals," which would serve as the fingerprint of extraterrestrial civilizations.

Natural astronomical objects also produce electromagnetic signals, but those are widespread and could be distinguished from artificially produced narrow bands. So far, no confirmed, artificially produced extraterrestrial signal has ever been found. The SETI@ home screensaver software asks the general public to contribute their free computing time to analyze the data SETI receives.

Some experts now caution us about the risk of contacting aliens, while others advocate a different approach in seeking aliens. Recently (2010), renowned British cosmologist Stephen Hawking warned that "if aliens ever visit us, I think the outcome would be much as when Christopher Columbus first landed in America, which didn't turn out very well for the Native Americans."

As we know, the laws of physics do not allow any travel beyond the speed of light, limiting the possibility of an alien visit from the distant corners of the universe. Even if alien civilizations could accomplish such high-speed travel, it must be at the expense of their enormous resources and time. What on Earth would make them to do that? Unless they have mastered laws of physics unknown to us, they won't be vacationing on this planet for fun.

In 2008, NASA broadcast the Beatles' song "Across the Universe" aimed at Polaris, the North Star. It would take 431 years for this beam, traveling at the speed of light, to reach the target location, which, in fact, is in our cosmic background. These attempts are more symbolic in nature and reveal the unfathomable nature of space and time that compose our universe.

Our radio transmission history is about 100 years old now. Our unintended radio waves have been propagating throughout space for about 100 years to alien worlds, if they exist. No one responded to our radio signals so far and is not expected to. Most of our radio transmissions would not be detectable with our current technology at far distances, say at the distance of the neighboring star. Also, signals have to be strong for detection at faraway places. Finally, in the time these signals have traveled into space the technology has undergone revolutionary changes.

What if the aliens are much more advanced than us? Say their civilization has been around for at least a million years more than ours. Then it's a different game. They might have a completely different mode of communication or transportation. If such aliens want to send us a message, then broadcasting may not be the best way for them to do that.

Some researchers suggest that they could have genetically engineered us in such a way that we transmit their message in a chemical form called DNA. Our bodies, which may be nothing more than a carrier of that message, then passes it on to the generations that follow. The non-coding DNA, which makes up about 95 % of the human genome, is called junk DNA, the functions of which are not completely understood. Some researchers suggest that the evolutionary traits in the junk DNA may help us to identify our alien connection.

For now, these suggestions are purely hypothetical in nature. As Carl Sagan noted, "Extraordinary claims require extraordinary evidence."

Paul Davies, the author of a public lecture called "The Eerie Silence: Renewing Our Search for Alien Intelligence" and director of the Beyond Center for Fundamental Concepts in Science at Arizona State University, has refuted Hawking's arguments in a

Figure 11.1. How would you discover Earth? Not by living here but by leaving it? It's a big blue marble or a pale blue dot, depending where you are looking at it from. The most important question is, are we alone or not? (Image credit: NASA).

Wall Street Journal blog. He says that a very advanced civilization may not be aiming for our resources, or even be interested in us; they might have mastered the know-how of extracting plentiful resources available elsewhere in the universe.

The question here is: Are we the result of a unique cosmic accident or just one among the many life-infected planets that share a common legacy? There is no guarantee that every intelligent civilization will survive for a long time, as they might destroy themselves or face the inevitable destruction caused by the cataclysmic events that are often unleashed in the cosmos. We don't even know how life began on this planet. How can we be sure that intelligent life exists and flourishes on another planet?

We may never find an alien civilization in our lifetime, or ever, but it is likely we will continue to investigate. So far, the universe hasn't disclosed all its secrets. But remember, our search is only decades old.

At this point, it is interesting to dive into a contemporary budget issue that limits our search for aliens.

SORRY ALIENS, WE HAVE A BIG DEFICIT

In another development, our longstanding desire to become part of a galactic empire by establishing contact with alien kingdoms has received a setback. The reason is, as one might expect, not the lack of aliens or their potential habitats in the universe but the cash crunch.

In 2011, the SETI Institute announced, "Federal and state funding cutbacks for operations of U. C. Berkeley's Hat Creek Radio Observatory (HCRO) force hibernation of Allen Array Telescope."

The Allen Array Telescope (AAT) opened in 2007 and is named after Microsoft co-founder Paul Allen for his donations to the project. It has been a dream come true for the SETI researchers. The Hat Creek facility needs $1.5 million a year for its operations and another $1 million for other costs. The radio telescopes at this facility scan signals from potential alien civilizations.

When it comes to SETI, everyone has either simple solutions or serious criticism, though this research has some ardent supporters. SETI astronomers had to do lot of public relations work in the past as well to keep going with their investigation. Even the U. S. Congress had slashed all the funding for SETI back in 1993, though that was lifted recently.

The goal of SETI is not to look for any primitive life forms, which might be plentiful even in our galactic neighborhood. SETI looks for developed civilizations, at least comparable to us or further advanced, that can communicate with radio signals the way we do.

Generally, people are tempted to think research efforts such as the particle hunt, definitely much more expensive, is something we cannot ignore and must be continued. However, one must admit that most of these experiments pursue particles that are theoretical in nature, including the God particle (Higgs boson). While these scientific pursuits are easily accepted and admired, the SETI, strangely, has a lower ranking in public perception.

Whether it is the alien hunt or the particle hunt, there is no doubt that serious science is invested in these efforts, and the spinoff benefits from these experiments fine-tune the technology we use now and would enable us to master new science.

A section of our populace, it appears, has developed some sort of resistance to any basic science research. Too many alien movies and popular UFOlogists have, perhaps, contributed to this indifferent attitude, and these folks, possibly, are doing their best to discredit serious scientific pursuits. It is time to reach out to other countries and engage them in collaborative large-scale alien search experiments similar to the LHC (Large Hadron Collider) experiment for economic and other reasons. This might ease our financial responsibility for such ventures, and when we finally contact the aliens, we could present ourselves as a single group to greet them.

The search for aliens is not often seen as a serious scientific pursuit for many reasons, including the influence of science fiction and the depiction of aliens. So let's look at some of the misconceptions about Center for SETI research, one of the centers under the umbrella organization, the SETI Institute.

SEVEN MYTHS ABOUT SETI

1. SETI detected extraterrestrial signals!!

   No confirmed extraterrestrial signal has ever been detected unambiguously. The "Wow" signal picked up in 1977 at the Ohio State radio observatory was something close to this in nature. But, it was never detected again and can't be considered as a scientific claim. Yet it remains in the history of alien search as the best candidate for an extraterrestrial signal.

2. We should pick up signals from an alien civilization very soon.

   Setting a time frame for such an experiment is a pure guess; even researchers are not sure about it. It could happen any moment or it may never happen. Some researchers (Loeb and Turner 2012) suggest that we should depend on different techniques such as looking for artificially illuminated objects in the outer Solar System and beyond rather than searching for traditional radio signals. This might accelerate our search.

3. Aliens are already among us.

   Unless we are talking about some sci-fi movies, this is absolutely false. The assumption that ancient civilizations and their structures are somehow linked to alien visits is a prevailing fallacy. Some believe there are traces of alien visits on Earth, such as crop circles. It is a possibility that on Earth there is more to life than meets the eye. The microbial organisms have been found to be thriving in extreme conditions, and some of them could be considered as aliens on earth. Other than that none of them are walking among us.

4. There is proof of UFOs and aliens.

   No credible evidence exists to suggest that people have seen UFOs and aliens. Many skeptics argue the other way, though. Some even claim that the abduction stories linked to aliens, which are filled with subjective narrations and untraceable facts, are true. However, researchers (Davis et al. 2013) say these events are more correlated to personality characteristics of people who report alien abduction experience.

5. Why don't we send signals to alien worlds?

   Researchers are not interested in it, primarily due to the distance involved. In a galaxy that is 100,000 light-years across, even if there is a technological society about 100 light-years away, it would take at least 200 years for us to get a reply, assuming they deciphered our message. However, some symbolic messages were sent in the past.

6. Our radio signals are currently being picked up by alien kingdoms.

   Some of our broadcasts may penetrate Earth's atmosphere and travel through space at the speed of light, but they would become scattered and less intense and thus difficult detect even with very sensitive and large radio telescopes. Hopefully, ETs are not listening to the funding cutback news, as they might lose any interest on a planet that has no money for intergalactic communication!!

7. It is a waste to engage in SETI research.

   Are we alone? We can't flee from that question whatever the outcome. The only way to find the answer is through investigations. Also, the spinoff benefits from any scientific experiment are valuable and help shape our society along with the better understanding of our universe.

Searching for extraterrestrial beings is, perhaps, just the culmination of human imagination and ingenuity. But, as the late Carl Sagan, a strong advocate of the scientific search for aliens said, "Imagination will often carry us to worlds that never were. But without it we go nowhere."

Our difficulties with space exploration are not just limited to SETI, as you might think. For decades, the space shuttle fleet was our transportation system to nearby space. Now they have become memories of the past.

SPACE SHUTTLES: THE GLORY OF THE PAST

On April 17, 2012, the space shuttle *Discovery* departed Kennedy Space Center atop a jumbo jet and headed to the Smithsonian Air and Space Museum's Udvar-Hazy Center, where it will become a museum highlight. For many, it's agonizing to watch these wonderful machines of yesteryear turn into mere museum exhibits. However, they bring back exciting memories of an adventurous era in space exploration spanning 30 years.

Over three decades, space shuttles have been America's flagship space program. Since the program's inception in 1981, shuttle launches became so usual that, after a while, not much attention has been paid to the event. The glorious vehicles that created history by carrying humans and cargo to space are becoming part of history.

Figure 11.2. Once a warrior of our space exploration program, *Discovery* is now doomed to be a museum relic. Atop a shuttle carrier aircraft, *Discovery* departs the Kennedy Space Center (Photo courtesy of NASA).

Although space shuttles inspired generations of space-gazers and contributed greatly to our national pride and identity, there are many misconceptions about the space shuttles and their missions. Let's look at those.

1. There were only five space shuttles.
   In fact, there were six space shuttles built, although one of them, as intended, never went to space. The *Enterprise* was the first space shuttle built, but its role was limited to tests to see how the orbiters would work. It was used in approach and landing tests, but never left Earth's atmosphere. The *Columbia* and *Challenger* were lost in accidents. That left three remaining shuttles – *Discovery, Endeavor,* and *Atlantis* – as survivors, though they will fly no more.
2. Shuttles can go deep into space.
   Not really. The shuttles were designed to travel in low Earth orbit. The orbital altitude of shuttles ranged from 190 miles to 330 miles, depending on the mission. However, the question of where space begins is a tricky one. About 60 miles above sea level is considered as the beginning of space, but different layers of Earth's atmosphere can extend much beyond that. There is no unanimous agreement on the boundary of space and Earth's atmosphere. The best answer is 'between here and there space begins,' depending who answers it. While the U. S. Air Force considers higher than 50 miles as space, the satellites must stay 200 miles and up to maintain permanent orbit around the earth.
3. People could have sex in the space shuttles.
   As one can expect, many stories and so-called experiments about this topic have been reported by the media to excite or confuse people. NASA vehemently denies any such experiments undertaken by the agency. It's definitely a struggle to even kiss in micro-gravity environment, let alone have sex, especially when there is no up and down. But the topic will continue to flourish in urban myths considering its innate nature. However, if long-term habitation in space is a priority, all biological needs, including sex in microgravity, must be a part of the whole picture.
4. The shuttle program was a waste of money.
   Some call the program a clunker. Humans have flown in space for about half a century. Billions of dollars have been spent. The benefits may not match the cost dollar-for-dollar. Many experiments have been conducted in space in basic physics, biology, and medicine and in several other areas. But the critics argue these could have been done in a cheaper way.

   However, NASA's annual budget is $18.7 billion through fiscal year 2016, which is much less than the federal government is spending on the Iraq war. The program played a big role in creating numerous technological products and made space tourism a reality. Moreover, the shuttle missions helped to build the International Space Station and launched the Hubble Space Telescope.
5. The space program will end with the space shuttles.
   Many obsolete technologies and inherent risks in the design of the shuttles, along with the relatively high cost of operation, are among the factors that grounded the shuttle fleet. By transferring the low-Earth services to the private sector, NASA can focus on more ambitious future missions to the Moon, Mars, and asteroids. The agency prefers

commercial U. S. firms to transport men and cargo to the International Space Station, but in the meantime it has to rely on Russian vehicles to do so. This is definitely the end of shuttle program but the beginning of, hopefully, more exciting space exploration.

One might wonder about the impact of the space shuttles now that they are part of our space history. The answer to that question can be paraphrased in a quote: "Dwelling on the past may be futile, but learning from it is worthwhile."

Now let's get back to our discussion on aliens. The stories about aliens are as old as our obsession with aliens.

OBSESSION WITH ALIENS

Our fascination with aliens is neither new nor will it fade away quickly.

The subject of aliens and alien invasions have been one of the most popular story lines in movies. The fact is that even before the Hollywood movies made it a profitable and popular topic, alien encounters and attacks were the purview of many well-known works that thrilled and delighted the general public.

For example, H. G. Wells' *The War of the Worlds*, published in 1898, describes the invasion of Earth by advanced aliens from Mars. As our rovers search the Martian surface for possible signs of life forms now, we know for sure that no such intelligent races existed, at least on the Red Planet. However, we can't deny the fact there is the possibility of some life forms surviving somewhere in this galaxy or beyond.

Some believe society will collapse in fear or chaos will ensue if we ever encounter a superior race of aliens. The claim that scientists and politicians will secretly engage or already has engaged in conspiratorial cover-ups about this topic is well circulated and believed by many, although such arguments have no foundations.

Even in recent times our fascination with aliens continues without waning. We could ask the reverse question – why don't aliens have an interest in contacting us? The feasibility of interstellar space travel could be a reason. The enormous amount of energy involved in the process, even for an advanced civilization, can be considered as a prohibiting factor to undertake such interstellar travel. Again, if such civilizations exist, their motivations and aspirations are more likely to be different than ours.

It has been estimated that the life on this planet originated about 3.5 billion years ago. There is some talk in scientific circles about the possibility of an extraterrestrial origin of life, though this view is not supported by many. This theory, known as panspermia, espouses the idea that the emergence of life on this planet is linked to an interstellar source and offers as evidence the fact that organic matter, the basis of earthbound life, is relatively common beyond the borders of this planet.

SHOULD WE COLONIZE SPACE?

Recently, a space venture project was announced by Planetary Resources (2013), a company whose investors include Larry Page, Google CEO, and Eric E. Schmidt, Executive Chairman of Google. The mission: harnessing valuable materials, anything from water to platinum, by mining the asteroids. This will provide, as the claim goes, stability on Earth, increase humanity's prosperity, and help establish and maintain a human presence in space. One can imagine the excitement over the announcement of this project. Coincidentally, the advisory board includes James Cameron, the acclaimed Hollywood director.

Although there is skepticism about the success of the project among some experts, many consider it as the inevitable part of human evolution and exploration. One big question arises, though. Do we want to use space commercially or preserve it?

The asteroid mining firm Planetary Resources has stated that its goal is to extract the raw materials from near-Earth asteroids. Recently, it has announced a plan to crowd-fund a project to build and operate a publicly accessible telescope. However, its primary focus is to hunt for asteroids orbiting near the Earth for resources. Robotic spacecraft will be used to mine the asteroids for precious metals and other materials.

Another company that has expressed interest in similar exploration is Deep Space Industries. It believes that the human race is now ready to begin harvesting the resources of space both for its own use and to increase the wealth and prosperity of all the people of planet Earth.

Asteroids are rocky objects, and they mostly exist in a region between Mars and Jupiter commonly known as Asteroid Belt. Scientists believe they are the leftover materials of our Solar System. These nomads often leave their home belt and move in our direction.

Figure 11.3. Artist's impression of an asteroid. New telescopes and technologies help astronomers to measure the sizes of small asteroids in the main belt for the first time (Image courtesy of ESO/L. Calcada).

In fact, many tiny asteroids enter our atmosphere every day, most of which burn up in the atmosphere or deposit very small pieces of themselves on Earth.

Mining the asteroids and building space colonies will have many technological challenges. The late Princeton physicist and space activist Gerald O'Neill not only envisioned the establishment of space colonies, but he published feasible plans and practical designs to accomplish such ventures with current technology. He asked, "Is the surface of a planet really the right place for expanding technological civilization?"

As an advancing civilization we will find new pastures, and there is nothing that stops us from doing so.

Searching for and extracting the resources and energy from beyond Earth is an indication that our civilization is moving from a lower to higher level. The Russian astrophysicist Nikolai Kardashev proposed a classification (1985) of civilizations based on their techniques of energy extraction. He categorized civilizations into three groups. A Type I civilization has the ability to harness all available energy sources on its own planet. As it progresses, it becomes a Type II civilization, which utilizes all the energy from its star. If this civilization survives a long time it will proceed and transform into Type III civilization. At this stage, it will depend other stars in their galaxy for their energy needs.

Although Kardashev's suggested classification scale is not to be taken as an absolute standard, the indications are that we are perhaps close to becoming a Type I civilization. This can be accomplished as long as we do not destroy ourselves.

The questions regarding jurisdiction over the asteroids and the financial implications of extracting resources are to be dealt with. However, those are no reasons to refrain from venturing into the future. Our civilization will continue its evolution.

While the asteroid mining is a plan that will unfold over long term, here we will briefly describe a mission named Stardust that was planned to visit a comet. The goal of this mission was to collect samples of the comet, called Wild 2, and return them to Earth for laboratory analysis. It was launched by NASA on February 7, 1999.

Although asteroids are solid objects and reside relatively close to us, the comets are mostly puffy bodies and travel to the outer edge of our Solar System. They are made up of a mixture of frozen water, gases and dust. When they approach the Sun, the heat from the Sun evaporates the material, which results in a long tail visible in the sky.

Scientists used aerogel collectors to trap samples from the coma and some interstellar dust. The precious cargo was brought to Earth in 2006 by parachuting a reentry capsule weighing approximately 125 pounds.

The material collected, very tiny grains even smaller than the width of a hair, was in fact the first sample of pristine comet material ever obtained. Scientists have discovered glycine, an amino acid that makes protein, in samples of Wild 2.

When Stardust was finished with its main task of collecting comet materials, it was redirected by NASA to do a different and final mission, where it had to fly by a comet named Temple 1 in 2011. Thereafter it was decided to put an end to the 12-year journey of the most famous comet hunter that encountered two comets and an asteroid during its billions of miles of travel.

Historically, comets are held in reverence and fear. They have the most remarkable appearance in the night sky. The ancient cultures described them as the "harbinger of doom" and "the menace of the universe." The ancients believed comets were the messengers of the Gods, and comets were often blamed for disasters on Earth.

Figure 11.4. Comet Wild 2 during the close approach phase of Stardust's Jan 2, 2004, flyby. The images show an intensely active surface, jetting dust and gas streams into space and leaving a trail millions of kilometers long (Image courtesy of NASA/JPL-Caltech).

Today, science and reason have proved that these are the distant cousins of our Earth and they are paying an occasional visit to display their ancient connection! The Stardust mission will be remembered as a steppingstone in our effort to excavate the objects beyond our own planet.

However, the threat of possible destruction caused by asteroids or comets is still real. On February 15, 2013, an asteroid hit the atmosphere above the Russian city of Chelyabinsk, causing many injuries and widespread damage. Events of this magnitude are expected once every several of tens to 100 years.

NASA has an active program to track near-Earth objects (NEOs). These are comets and asteroids that have entered into a danger zone where a potential collision with Earth is possible. The goal of this program is to detect, track, and characterize these objects that could approach Earth. So far, this program has identified 10,000 such objects.

Over the centuries, the human perception of comets has changed dramatically –from messengers of God to nomads of the Sun's family. Although we view them differently than ancient people did, a few researchers believe that the comets carry some very special cargo. They even claim the evidence for that, as the following incident describes.

Figure 11.5. A small asteroid entered Earth's atmosphere over Russia and fragmented at high altitude, creating a fireball with dazzling light (Photo courtesy of Uragan. T T/Wikimedia Commons).

WHEN THE RAIN GOD CRIED RED

For the southern Indian state of Kerala, red is not truly an unfamiliar color. The state's powerful leftist organizations often garnish the streets with red flags and signs in an apparent show of strength. But when the red rain arrived in July 2001, it caught everyone off guard. Now, even after a decade, its implications are debated in scientific circles. Some researchers suggest this phenomenon was an obvious mark of an alien life form.

During that time, sporadic rain brought spells of colored showers in many parts of the state, mostly scarlet water. The dust particles in the atmosphere could color the rain – this was the most reasonable explanation initially. But the particles showed biological cell-like structure under the microscope, or so goes the claim.

Godfrey Louis, now a physicist at Cochin University of Science and Technology examined the red particles collected from different parts of the state. He tentatively claimed that the absence of DNA in the samples was an indication of the alien origin of these microbes. Seemingly, they had fallen to Earth on a comet that might have exploded above Earth before the red rain. The results were published in 2006.

Later on, the samples were shared with Prof. Wickramasinghe of Cardiff University in the UK, a well-known astrobiologist who worked with the late Fed Hoyle, a champion of the panspermia theory. Panspermia suggests that organic materials or even simple life forms exist all over the universe in comets and interstellar dust clouds and are being transported from one location to another. Microorganisms remain frozen, and a small fraction of them could survive the extreme conditions to reach other parts of the cosmos. Accordingly, life on Earth was most probably seeded from one of these alien sources. In other words, our ancestral origin is extraterrestrial.

The subsequent investigations of this rain sample using different techniques concluded that the red rain particles were an unusual type of biological cell. Wickramasinghe agreed that the cells do have properties that are not shared by other microbial cells on Earth. Although that does not inevitably make them alien, there are indications that they may be so.

In a 2010 pre-print paper (Gangappa et al. 2010), Wickramasinge and other authors described that the cells found in the red rain survive and grow after incubation for periods of up to 2 h at 121°C. Under these conditions, daughter cells appear within the original mother cells, and the number of cells in the samples increases with length of exposure to 121°C. No such increase in cells occurs at room temperature, suggesting that the increase in daughter cells is brought about by exposure of the red rain cells to high temperatures.

The widely accepted view has always favored a terrestrial origin of life. Life began on Earth in a primordial soup in the simplest possible way about 3.5 billion years ago. But, the proponents of the panspermia theory argue that life is a cosmic phenomenon, not just a local incident associated with this planet. As one can imagine, it is not easy to get a warm reception for such a theory because science demands extraordinary evidence for extraordinary claims.

In the meantime, in Kerala, several monsoons have come and gone, many water wells have vanished, some rivers have died, and the streets became red again. People forgot about the red rain as rain and water became scarce. Also, the alien seeds never came back riding on the comets.

To conclude, all known cultures that ever existed had their own version of alien stories that show alien beings ranging from winged angels to little green men. But, for scientists no such descriptions can be made until we have clear evidence. Should we search for it? Yes. Is it time to make definite conclusions? No.

WILL WE FIND ALIENS?

Our notions about aliens are shaped mostly by the popular culture. Science fiction works and movies contributed to our lasting memories of little green men. However, in the scientific point of view, aliens do not necessarily mean an advanced civilization whose subjects are wielding laser swords.

In that sense, our best chance of finding alien life forms might come from the Curiosity rover, currently in action on Mars. Although we do not expect a complex biosphere such as on Earth, there are indications that Mars might have harbored some form of life in the past.

Figure 11.6. This artist concept features Curiosity's Mars Science Laboratory along with an illustrated astronaut bird (Image courtesy of NASA/JPL-Caltech).

Figure 11.7. In the search for distant worlds, telescopes such as ESO's 3.6-m telescope and the Swiss 1.2-m Leonhard Euler Telescope will continue to search for planets beyond our Solar System (Image courtesy of ESO).

The Curiosity rover with its Mars Science Lab has the ability to analyze samples on the surface of Mars. Though it is a descendant of previous rovers, such as Spirit and Opportunity, Curiosity is much more advanced than any other rover landed on an astronomical body.

Curiosity has already revolutionized our notion of the Red Planet. NASA scientists announced in March that the planet once could have supported life after analyzing rock samples collected by the rover. Curiosity's activity on Mars, using an array of science instruments, began only about a year ago. Another year of prime research is awaited from Curiosity, and astronomers are optimistic about Curiosity's ability to write new Martian Chronicles!

So, in the short-term, if we discover alien life forms once existed on Mars, even if they are some sort of primitive microbes, that alone will be a significant find. On the other hand, beyond our Solar System, with our advanced telescopes and techniques, we should eventually be able to peer into the atmosphere of newly found planets and be able to confirm the signature of life, if it is flourishing on those planets.

In the meantime, we could assume that the advanced aliens are contemplating their life in their own abodes and might have ascended to a divine level, losing interest in any material worlds such as our little blue planet.

REFERENCES

BBC News – Stephen Hawking warns over making contact with aliens. Available at http://news.bbc. co.uk/2/hi/uk_news/8642558.stm. Accessed 12 May 2010.

Davies, P. (2010). *The eerie silence: Renewing our search for alien intelligence* (p. 4). Boston/ New York: Houghton Mifflin Harcourt.

Davies, P. (2011). *The eerie silence: Renewing our search for alien intelligence* (1st ed.). Boston: Mariner Books.

Davis, T., Donderi, D., & Hopkins, B. (2013). The UFO abduction syndrome. *Journal Of Scientific Exploration, 27*(1), 25–42, Academic Search Premier, EBSCOhost, viewed 27 March 2013.

Deep Space Industries – MISSION. (2013). Deep space industries – Mission. Available at http:// deepspaceindustries.com/mission/. Accessed 25 Jan 2013.

deGrasse Tyson, N. (2007). *Death by black hole: And other cosmic quandaries.* New York: W. W Norton & Company.

Drake Equation|SETI Institute. (2012). Drake Equation|SETI Institute. Available at http://www.seti. org/drakeequation. Accessed 24 Nov 2012.

Dyson, F. J. (1960). Search for artificial stellar sources of infra red radiation. *Science, 131,* 1667–1668.

Elsila, J. E., Glavin, D. P., & Dworkin, J. P. (2009). Cometary glycine detected in samples returned by Stardust. *Meteoritics &Planetary Science, 44*, 1323–1330.

Gangappa, R., et al. (2010). *Growth and replication of red rain cells at 121 degree Celsius and their red fluorescence.* http://arxiv.org/abs/1008.4960. Accessed Nov 2011.

Gray, R. (2012). The elusive wow!. *Sky & Telescope, 124*(4), 86, Academic Search Premier, EBSCOhost, viewed 27 June 2013.

Ishii, H., Bradley, J. P., Dai, Z. R., Chi, M., Kearsley, A. T., Burchell, M. J., Browning, N. D., & Molster, F. J. (2009). Comparison of comet 81P/Wild 2 dust with interplanetary dust from comets. *Science, 319*, 447–450.

Kardashev, N. S. (1985). On the inevitability and the possible structure of supercivilizations. In *The search for extraterrestrial life: Recent developments*, IAU Symposium 112, pp. 497–504.

Kardashev, N. S. (1997). Cosmology and civilizations. *Astrophysics and Space Science, 252*, 25–40.

Loeb. A., & Turner, E. (2012). Detection technique for artificially-illuminated objects in the outer solar system and beyond, earth and planetary astrophysics. Available at http://arxiv.org/abs/1110.6181. Accessed 20 Apr 2013.

Mars Science Laboratory. (2013). Mars science laboratory. Available at http://mars.jpl.nasa.gov/msl/. Accessed 10 Mar 2013.

Nadis, S. (2010). How many civilizations lurk in the cosmos? (Cover story). *Astronomy, 38*(4), 24–29, Academic Search Premier, EBSCOhost, viewed 27 April 2013.

NASA – NASA and The Beatles Celebrate Anniversaries by Beaming Song 'Across The Universe' Into Deep Space. (2011). NASA – NASA and The Beatles Celebrate Anniversaries by Beaming Song 'Across The Universe' Into Deep Space. Available at http://www.nasa.gov/home/hqnews/2008/jan/HQ_08032_NASA_Beatles.html. Accessed 2 Sept 2011.

O'Neill, G. K. (1975). Space colonies and energy supply to the earth. *Science, 190*(4218), 943–947.

O'Neill, G. K. (2000). *The high frontier: Human colonies in space: Apogee books space series 12* (3rd ed.). Burlington, Ontario, Canada: Collector's Guide Publishing.

Planetary Resources – The Asteroid Mining Company. (2013). Planetary Resources – The Asteroid Mining Company. Available at http://www.planetaryresources.com/. Accessed 25 June 2013.

Sagan, C. (1980). *Cosmos*. New York: Random House.

Sagan, C. (1994). *Pale blue dot: A vision of the human future in space*. New York: Random House.

Sagan, C. (2000). *Carl Sagan's cosmic connection: An extraterrestrial perspective*. Cambridge: Cambridge University Press.

Schwartz, R. N., & Townes, C. H. (1961). Interstellar and interplanetary communication by optical masers. *Nature, 190*, 205–208.

Tarter, J. (2001). The search for extraterrestrial intelligence. *Annual Review of Astronomy and Astrophysics, 39*, 511–548.

Stewart B. (1976). Is the surface of a planet really the right place for expanding technological civilization? *Interview 1975*. Available at http://settlement.arc.nasa.gov/CoEvolutionBook/Interview.HTML. Accessed 12 June 2012.

12

The End of Everything

"I had a dream, which was not all a dream.
The bright sun was extinguished, and the stars
Did wander darkling in the eternal space,
Rayless, and pathless, and the icy earth
Swung blind and blackening in the moonless air."

– Lord Byron, English poet of the Romantic Period
(Excerpt taken from the poem "Darkness")

People often talk about the end of the world. Back in the 1980s, a pastor in the author's hometown would always refer to this topic. He would quickly find Bible verses to prove his point. A torrential rain during the monsoon season, which is not uncommon, would usually flood the river that flows through our town. The pastor would find an opportunity there to link to Noah's Ark and the Great Flood.

This is widespread thinking in many cultures and many parts of the world, to be alarmed about the end of the world. Most people thought their generation would be the last to survive, and the end was near.

Over the past millennia, various spiritual groups have prophesied the end of the world. The chain of cataclysmic events, such as earthquakes and tsunamis, often feed apprehensions of the end of days. Many religious texts and ancient scriptures give warnings about the end of the world. One such eschatological belief, attributed to the Mayans in contemporary times, postulated that the world would end in December 2012.

None of the end of the world prophecies has come to pass. But the end of our universe is inevitable, according to cosmologists. Fear not, though, because the end will not come for trillions and trillions of years. And when it does occur, underground bunkers will do you no good, as it would involve the end of space, time, matter, and energy – the ultimate destruction of everything.

S. Mathew, *Essays on the Frontiers of Modern Astrophysics and Cosmology*, Springer Praxis Books, DOI 10.1007/978-3-319-01887-4_12, © Springer International Publishing Switzerland 2014

Figure 12.1. Looking where earth and the heavens meet (Image credit: CC by Hugo Heikenwaelder on Wikimedia).

Cosmologists debate over the diverse ways our universe could end, but there is no doubt in their minds about the ultimate destiny of the cosmos. In other words, there is no "forever." The eternity we describe is a borrowed concept granted mercifully by the forces that govern the cosmos.

In cosmology, the end of the universe is as inevitable and inescapable as the laws of nature that predicted its birth. The cosmologists' vision of the ultimate end is markedly different from the scenarios that we can contemplate and are much more complex than we can imagine.

The notion of the end inspired poets such as Robert Frost, who wrote in his poem, "Fire and Ice":

Some say the world will end in fire,
Some say in ice.
From what I've tasted of desire
I hold with those who favor fire.
But if it had to perish twice,
I think I know enough of hate
To say that for destruction ice
Is also great
And would suffice.

Fire and ice, in fact, are two cosmological models of the world's end. Of the numerous depictions about the fate of the universe, the most commonly accepted one is the Big Freeze, in which the universe will continue to expand forever, making it a cold and isolated place until the temperature reaches absolute zero. Beyond that point, molecular motion ceases and hence the existence of everything we know.

Before the Big Freeze, physicists contemplated a fiery destruction of the universe. If the universe is a closed system, eventually the heat energy of the system, from sources such as stars, will be dissipated uniformly throughout the universe. Entropy, a measure of disorder, must increase in a closed system under the laws of physics. This will bring an end to any kind of motion or physical phenomena in the cosmos as well as any form of life. But this mode of death for the cosmos lost ground to the Big Freeze when astronomers realized that the universe is expanding at an accelerated pace.

In the Big Crunch scenario, in the far future the universe will stop expanding and collapse into itself. All the matter will be pulled in and turn into a black hole.

Apocalyptic predictions for Earth and its inhabitants have been plentiful through the ages. For example, Biblical references to the end, though mostly metaphorical in nature, are major elements of the Old and New Testaments of the Bible. The Book of Revelations in the New Testament and the Book of Daniel in the Old Testament are descriptive of the disasters that await Earth and humans. However, these revelations are mostly geocentric; it is no surprise that neo-prophets and doomsday-sayers draw their energy from these Biblical passages. Social upheavals and natural disasters have been a feeding ground for a generation of forecasters.

Even our brightest minds were not completely free from the end-of-world prophecies…

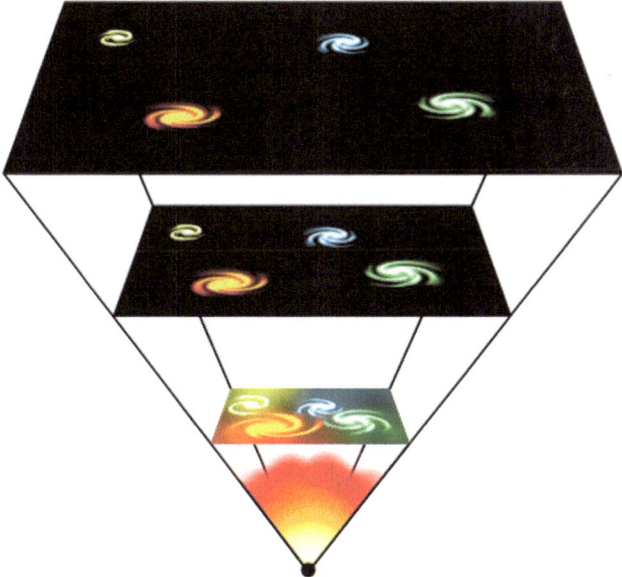

Figure 12.2. The end of our universe will happen in trillions of years from now. How will our universe end? The Big Crunch is one scenario (Image courtesy of Bjarmason/Wikimedia commons).

WORLD TO END IN 2060?

It may seem paradoxical that the father of modern science, Isaac Newton, is considered to have calculated the time for the end of the world based on verses from the Book of Daniel. However, it is not surprising, considering his Christian faith, that he searched for wisdom in biblical codes.

Newton's theological and mystical writings were on display at an exhibition entitled "Newton's Secrets," at the National Library of the Hebrew University of Jerusalem in 2007. It was the first time that Newton's theological manuscripts, in Israel since 1969, were presented to the public. In one of them, Newton calculated that the world will end in 2060, based on a phrase from Daniel 12:7 "for a time, times, and a half." Newton interpreted this phrase as meaning 1,260 years would pass from the establishment of the Holy Roman Empire by Charlemagne in 800 until the end. It is not clear whether this was an attempt by Newton to put an end to, in his own words, the rash conjectures of fanciful men who frequently predicted the end time, and by so doing discredited sacred prophesies, which commonly fail.

For modern-day scientists, many causes could bring an end to this dynamic planet and the life that it harbors. A huge asteroid impact, a super volcano, or manmade disasters, such as global warming or nuclear war, are all capable of creating global calamity. But, suggesting a particular date for such events would be outside the realm of science. The laws of nature that enable us to predict the exact date of tides and eclipses for decades to come do not suggest anything that can have a catastrophic effect on this planet. Yet, the potential disasters remain a possibility in the grand cosmic scheme, where this small rocky planet and its masters are an insignificant breed of chemical beings that could easily be terminated.

The so-called Judgment Day and the end of the Mayan calendar, which have been popularized recently, lack any scientific basis. Spiritualists have bet on many dates in the past, and they were off base this time around as well.

Another example of an end of the world scenario recently surfaced in the media. Let's take a look at it.

THE MOON IS NOT A HARSH MISTRESS

The speculation that the full Moon and its gravitational effect on Earth could trigger earthquakes and other natural disasters is getting attention in some quarters. It is not surprising, given the fact that humans have been looking above rather than below our feet for answers since ancient times.

March 19, 2011, was a full Moon day, in fact, a special one dubbed as "supermoon." On this day, the Moon was closer to Earth than it had been in its last 18 years – full and close. It approached Earth at a distance of 221,567 miles – lunar perigee.

The gravitational pull of the Moon on Earth is evident from the ocean tides. Basic science tells us that the tides are caused by the varying gravitational pull of the Moon across Earth. Stronger tides near the full Moon are nothing unusual. It has been known that when the Sun, Moon, and Earth are along a straight line, as in the case of a full Moon, the tidal effects are increased.

Figure 12.3. Supermoon rising near Lincoln Memorial on March 19, 2011, in Washington, DC (Photo courtesy of NASA/Bill Ingalls).

Although the Sun has a larger gravitational influence on Earth, the difference in the Sun's gravitational force between one end of the Earth and the other is not as much as that of the Moon's – the reason tides are caused mainly by Moon's gravity.

The whole idea that the proximity of the Moon could affect Earth seems to have some weight. It is true that Moon's gravity can cause land tides similar to ocean tides. But that this stress can accelerate tectonic activity to cause an earthquake is beyond any known science now. We need to know the real science rather than mere coincidences or the statistical flukes.

The recent Japanese earthquake and the full Moon had no correlation at all. The Moon was at its perigee on March 19 and was actually farther than the average distance on the day the Japanese quake hit.

Scientists have studied the Moon for decades and have found no conclusive evidence to connect the Moon with seismic activities here on Earth. Being in an elliptical orbit, the Moon's distance varies as it orbits Earth. It will go through perigees and apogees several times a year. On some of these occasions we may face natural disasters, but many other days are just fine. If we insist on finding a correlation between Earthly events and celestial phenomena, we could find plenty of them, and there are numerous such incidents in the universe to link even with our everyday activities. No wonder that in the past (perhaps even now) people believe the Moon can affect our individual moods. Remember the phrase lunatic!

If we are determined to search for the root cause of earthquakes, it may be a better idea to look for reasons below Earth's surface. There are currently debates about the correlation between drilling and the possibility of earthquakes. There could also be several other factors, which we don't know yet, natural or manmade, that trigger earthquakes. By all accounts, the Moon seems to be a minor player now.

The Moon's gravitational pull on Earth at lunar perigee is not hugely different from other times, which can be calculated from simple mathematical equations. Even if we consider the effect of the Sun along with that of the Moon when they align together, it is not significant enough to alter the internal balance of Earth. The only outcome on March 19 was that of an apparently bigger Moon that offered a special treat for sky gazers.

In Hindu mythology the Moon God known as Chandra is identified with the deity Soma. He rides through the sky in a chariot drawn by white horses and antelopes. Soma was also the legendary elixir of immortality that the Gods drink. The myths say the Moon is filled with the elixir. It is a strange coincidence, then, that the Indian Space Research Organization's Chandrayaan-1 mission has helped to identify water – the elixir of life – on the Moon. The detection of water on the Moon was made possible by NASA's Moon Mineralogy Mapper (M3) aboard ISRO's Chandrayaan-1 spacecraft.

What drives such explorations? The Moon, of course, our closest celestial body, has been an integral part of our culture. Perhaps it generated more inquisitiveness in our mind than any other objects in the sky. However, our desire to know is the bigger driving force than the rocket engines that reach the Moon. Such desires are expressed even in ancient texts, though they often describe celestial bodies as Gods who dictate our lives.

O Moon! We should able to know you through our intellect. You enlighten us through the right path. (Rigveda, Part 1/91/1)

Chandrayaan-1, India's first mission to the Moon, was launched successfully on Oct. 22, 2008. The spacecraft orbited around the Moon at a height of 100 km from the lunar surface, looking for chemical, mineralogical, and photogeologic mapping of the Moon. The spacecraft carried many scientific payloads built in India, the United States, and Europe. The discovery of water – in small amounts but considered to be widespread on the surface of the Moon – stands as one of the most significant findings in modern planetary science.

When Apollo astronauts first brought Moon rocks home, there were some early suspicions of water in the samples. However, that was explained away later as possible contamination after landing back on Earth because the sample containers were found to be leaky. Thus, the Moon has long been thought to be a dead, dry world – a place, in the words of astronaut Buzz Aldrin, of "magnificent desolation." But some scientists suggested that water-ice millions of years old might be found in the shadowed craters of the Moon's north and south poles, where the sunlight does not reach. Analysis of data from the Moon probe showed that the absorption of wavelengths of light was consistent with absorption patterns of water molecules.

Regarding the new discovery, researchers think that the water is created when hydrogen atoms carried by the solar wind slam into oxygen-rich materials on the lunar surface, combining to form hydroxyl and water.

Besides NASA's M3 probe, India's Moon Impact Probe (MIP) also picked up signals confirming water on the Moon. This newly confirmed presence of water could 1 day help sustain lunar astronauts for longer or permanent stays there. It may take decades of technological advancement to harvest that water, though.

Another story that propagates the notion of an end scenario has been attributed to imaginary planets that exist.

EVEN THE WISE CANNOT SEE ALL ENDS

Nibiru, Nemesis, and now Tyche – the list of fictional or hypothetical astronomical bodies waiting to unload their vengeance on our little blue planet continues to grow, at least some believe so. Since time immemorial celestial bodies are known to have the liberty to alter our destiny, and much hasn't changed even today.

Nibiru has its origins in possibly the Sumerian fables. It is often referred to as Planet X, though this is a very general term. It's pure myth, but the Internet breathed a new life into this hoax and helped it to blossom. The doomsday prophets predicted that Nibiru was headed toward Earth with a possible collision in 2012 to coincide with the Mayan long count calendar. It is true that astronomers have discovered several dwarf planets in recent years, and there might be many more at the fringes of our Solar System. However these bodies stay in their own orbits that will never bring them near Earth. There is no evidence that supports a planet like Nibiru, but it may continue to exist in web pages and chat rooms. The best the doomsday advocates could do is to move the collision date beyond 2012 to keep the story alive.

Nemesis had its own nemesis years ago. This hypothetical object was proposed back in the 1980s as a companion to our Sun. This dwarf star was supposed to follow an elliptical path, perturbing comets in the Oort Cloud roughly every 26 million years. This would send a barrage of comets in our direction, causing terrible events that could eventually lead to mass extinction. Recent scientific scrutiny no longer supports the statistical periodicity of mass extinction and negates the Nemesis hypothesis. It is true that our planet was subjected to bombardment of comets and asteroids in its history, with major hits in every million years. That possibility still exists, but our ability to foresee such a scenario has increased several folds in modern times.

The latest hypothetical object Tyche was based on a mathematical model created by John Matese and Daniel Whitmire, researchers at the University of Louisiana at Lafayette. Their published paper proposed a companion for our Sun about 1–4 times the mass of Jupiter. It is a possible scenario in which our Sun will have a yet to be seen companion with an orbital period of a few million years. Again, that doesn't mean devastation for this planet. In Greek mythology Tyche is the benevolent sister of Nemesis, the Greek Goddess of retribution.

These findings are based on the study of the deviation of comet paths in the Oort Cloud. Their paper expresses the hope that the Wide-field Infrared Survey Explorer (WISE) might have recorded information about Tyche, which would be released in coming months. Though it is premature to say whether WISE data confirms or rules out a large object in the Oort Cloud, it has been generally agreed that WISE is capable of detecting all major objects in the infrared spectrum such as Tyche. According to the Jet Propulsion Laboratory, it is likely but not a foregone conclusion. The WISE mission has recently completed the sky scan and is in hibernation mode now.

Just as our ears cannot hear all wavelengths of sound and our eyes cannot see all wavelengths of light, our machines cannot see everything. However, finding new large heavenly bodies, even if they exist, as dangers lurking in the shadows of deep space, have no little support. Ultimately, the peril could be from super volcanoes to superbugs, or even from our own actions.

Even with the scientific advancements in all fields our irrational fear give rise to unwarranted responses some time. The section below describes a disaster and its aftereffects.

THE CHAIN REACTION OF A DISASTER

The 2011 Japanese earthquake and the resulting tsunami caused a renewed "nuclear fear." Although there is a need to address and review safety concerns and preparedness periodically, panic and overreactions are rooted in irrational fear.

We are subjected to a small amount of nuclear radiation every day, which varies depending on where we live. It's all around us in the air, water, and soil and has been since the formation of Earth. We can get it from radon gas that enters our house or from medical scanners.

This is no reason to suggest that we should ignore radiation effects. Health physicists mainly consider three kinds of exposure – inhalation, ingestion, and direct exposure. Nor do we intend to say that we shouldn't prepare for an accident such as the one that happened in the Japanese nuclear plants. Radiation can cause cancerous and non-cancerous health effects, which have been studied intensively. At the same time, the preparations and precautions must be based on facts, not just the information that comes in the radio waves that fill our rooms.

The metric unit of radiation doses is Sieverts (1,000 mSv = 1 Sv). According to the Nuclear Regulatory Commission (NRC) in the United States, the average person is exposed to about 6.2 millisieverts a year, mostly from background radiation and medical tests. (See chart.)

Although Colorado may have relatively higher radiation levels, places like Guarapari (24.5 mSv/year) in Brazil and Kerala (15.7 mSv/year) in India are considered to be the places where high natural background radiations are present due to the presence of monazite sand. But no statistically significant biological effects have been reported in the study, according to the International Atomic Energy Agency.

It is possible that granite countertops contain traces of naturally occurring radioactive elements that can emit radiation equivalent to the amounts emitted by some smoke detectors. The effects of low-level radiation is so insignificant, they may not even be detectable. And, if you are in space, you are exposed to cosmic radiations and its secondary radiations as well.

In many places in Japan the measured level of radiation is much less than that of a typical CT scan (about 10 mSV), though it reached much higher levels near the plant

Sources of Radiation

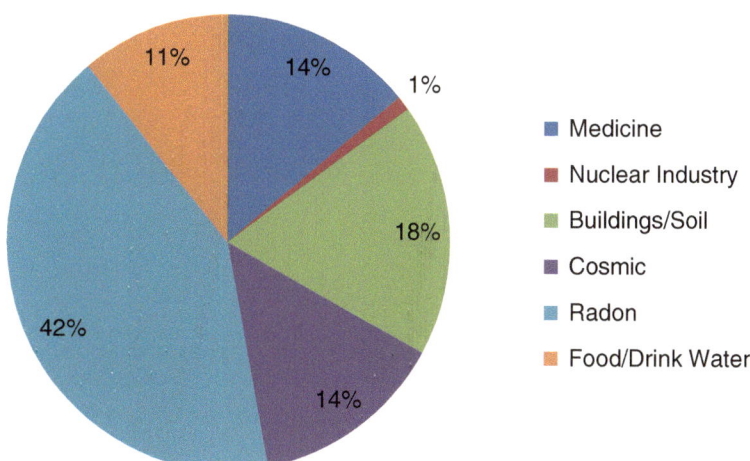

Figure 12.4. Sources of radiation (Credit: World Nuclear Association).

(about 400 mSv/h). The prolonged exposure time could add to the overall amount of radiation, especially with gamma and X-rays, which are the primary concern for external exposure. But when the radioactive material is inside the body, the exposure time is determined by its half-life. The alpha and beta particles are the main concern for internal exposure.

Radiation decreases more rapidly with distance by a factor of the square of the distance from the source. The level of radioactivity that has already been released in Japan would be in extremely low levels by the time it reaches the west coast of the United States – according to a unanimous panel of experts.

Industrial accidents, in the oil and coal sectors, are many times worse than nuclear power accidents, including the Chernobyl incident. The sudden failure of a hydroelectric dam would be more catastrophic for humans.

It is true that there are challenges involving the isolation and disposal of radioactive waste. The industry surely must maintain the highest levels of safety standards. But it's not the time to engage in conversations about abandoning nuclear energy. At least for now, there is no anti-radiation vaccine available, but we can check the facts and act accordingly instead of pressing the panic button.

To summarize, here is a 1953 quote by Albert Einstein: "The discovery of nuclear chain reactions need not bring about the destruction of mankind any more than did the discovery of matches. We only must do everything in our power to safeguard against its abuse."

It has been about 60 years since Einstein spoke those words; however, debate continues to rage. Nuclear energy is and will remain a major source of our energy needs. Obviously, we are responsible for ensuring that nuclear energy is handled safely. But the solution is

Figure 12.5. The nuclear fission of uranium is represented in this schematic diagram (CC Wikimedia Commons).

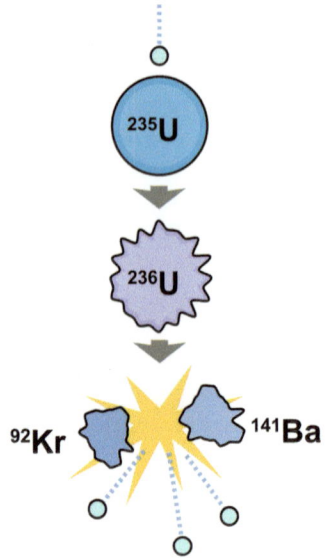

not to forbid its use. As worldwide energy consumption rises, countries will face serious challenges to meet their energy needs. Alternate energy sources have the capacity to partially fill the needs along with nuclear power, but not by themselves. Nuclear fusion is still in the research stage, and when it is developed we can come to depend on it as more efficient and cleaner than nuclear fission energy.

In Chapter 6, we discussed nuclear fusion and its possibilities. To complete the topic, we will briefly mention nuclear fission as well. While nuclear fusion is the union of lighter elements to form heavier ones, nuclear fission is the splitting of larger nuclei such as uranium to form smaller ones, along with energy release.

In the case of fission, a neutron smashing the uranium atom breaks it into two smaller elements (barium and krypton). The newly generated neutrons will further split more uranium atoms into smaller components, and a chain reaction occurs. This process will eventually split most of the uranium atoms to generate a large amount of energy.

When we talk about nuclear plants or nuclear bombs we are referring to nuclear fission, not fusion. As just mentioned nuclear fusion is yet to be realized though uncontrolled nuclear fusion is the science behind hydrogen bombs.

By studying the smallest of the small, we have learned to control the atoms and their phenomena. Apparently, we have the ability to use them safely.

Although we find the popular end of world scenarios are trivial, cosmologists view creation and destruction in the universe as cyclic phenomena.

CYCLES OF CREATION AND DESTRUCTION

For cosmologists, the fate of this planet and its Solar System is just one item on their menu of colossal ends. It has been estimated that Earth's continents will collide in 250 million years, and the Sun will fade away as a white dwarf in about 7 billion years. This 'blue dot' we know as Earth will be engulfed by the Sun well before that. The fate of the universe, on the other hand, evolves on a larger time scale – in trillions of years.

However, if the expansion of the universe continues at an accelerated pace, as it is doing now, it will be torn apart by dark energy at some point. From majestic galaxies to paltry subatomic particles, everything will be shredded by this phantom energy. The universe will be a great void caused by the Big Rip, another doomsday scenario predicted for the cosmos.

Death by ice or fire seems like a classical choice – simple and elegant. But the current understanding of the universe offers more possibilities. The theory branded as the Big Crunch is exactly the opposite end of the symmetric picture generated by the Big Bang. In the far future, the universe will stop expanding and begin contracting. All the matter and even space-time will collapse into a singularity, similar to what existed before the Big Bang. Some researchers, such as Martin Bojowald of Pennsylvania State University, believe that the Big Crunch will create a repulsive gravitational force that will result in a Big Bounce, which in turn will be the springboard for another universe (Bojowald 2006). This oscillating cycle of Big Bangs and Big Crunches will repeat in sequence. In that sense, one could argue the universe is infinite and eternal.

Lately, a growing number of researchers consider the cosmic cycles of death and rebirth as another possibility, a variation of oscillating cycles. The Hindu cosmological picture, which asserts the cyclical nature of the universe, is not very optimistic for those who seek eternity. Hindu cosmology posits that the world passes through four Yugas – Satya Yuga, Treta Yuga, Dvapara Yuga, and Kali Yuga. Based on the ancient Hindu times-cales, we are currently 5,000 years into Kali Yuga, which will last a total of 432,000 years. In this Yuga, infamously popular as the last Yuga, ignorance and evil grow ever greater in society. Finally, the tenth avatar of Vishnu, Kalkin, will put an end to evil at the end of this age. It is predicted that at the end of the Kali Yuga, the universe will be recycled by the Lord of Destruction, Shiva. After the dissolution, Lord Brahma will recreate the universe, everything will repeat, and the cycles will go on.

Physicists predict that trillions of years into the future, the information that currently allows us to model how the universe expands will vanish from the visible horizon as galaxies drift away from each other, making it impossible for light to reach beyond their horizons. What remains will be "an island universe," with the Milky Way and its local neighbors in an overwhelmingly dark void.

Our descendants, looking at the cosmos might be inclined to say that the Milky Way is the only galaxy and is the whole universe (unless, of course, they are taught otherwise). Incredibly, that is exactly what our ancestors thought. However, it is not just the universe that is evolving, but our intelligence and life itself. We cannot foresee how future generations will perceive the universe. They might not be interested in this cosmic illusion, as in trillions of years from now they may have transformed into beings that lose the desire to know anything, and the great Maya[1] dissolves in Brahman.

As mentioned earlier, the neo-prophets and doomsday-criers generally draw their energy from biblical passages or similar ancient texts, adding their own interpretations.

---

[1] The ultimate Illusion.

As Einstein wrote, "The world is a dangerous place, not because of those who do evil, but because of those who look on and do nothing." Or, maybe due to folks just waiting around to see the end.

All the above scenarios and discussions converge around a single question – can we predict the future?

PREDICTING THE FUTURE

In many cultures, the stars and planets play a major role in predicting the future of humans and their world. Now astronomers, with the help of the Hubble Space Telescope, say they can predict the future of the stars – not just for a few years but for at least 10,000 years into the future.

We are so used to looking at the past that we don't even realize it, and, rather strangely, we deem it as the present. When we look at the Sun, we see it as it existed about 8 min ago, for the finite speed of light, though it is a mind-blowing 186,000 miles per second, takes that much time to reach the 93 million miles to Earth. In a similar fashion, an object at a million light-years away could be seen with a telescope. Again, that picture represents the object's state a million years ago. Astronomers are particularly familiar with this issue and have no choice but to live with it.

The researchers used the Hubble telescope to map the motion of 100,000 stars in the Omega Centauri globular cluster over a period of 4 years. Omega Centauri is one of the star clusters in the Milky Way Galaxy and is about 15,000 light-years from Earth. This star cluster could often be seen in the southern sky with unaided eye. In fact, the ancient astronomer Ptolemy even cataloged this as a single star. Later on, it was identified as a behemoth cluster of stars. And now, with Hubble's sharp-vision cameras, astronomers know what Ptolemy thought as a single star is a compact cluster of millions of stars moving about like a swarm of bees. The Milky Way contains about 150 such star clusters.

Employing the images that were taken in 2002 and 2006, astronomers created a simulation of the motion of the cluster's stars. This would represent the future motion of the stars projected over the next 10,000 years. Assuming the simulation to be an accurate depiction of the cluster based on the known laws of nature, such as gravity and speed of light, strangely, this future vision would represent the state of the cluster as it existed about 5,000 years ago.

Although in modern time we use scientific observations for future predictions the medieval world had not distinguished astrology and astronomy, and its relics are still seen in our popular culture.

ASTROLOGY AND ASTRONOMY

Astrology and astronomy, not estranged in the early days of science, are clearly distinguished these days, with astrology considered as a pseudoscience, although some would disagree. Astronomy, on the other hand, has the strength of observation and rational principles to support it. Armed with such tools, modern-day astronomers take pride in predicting the future of stars or even the universe. Interestingly, such predictions are the norm in astrology, which are based on feeble myths and supernatural stories.

What enable astronomers to look into the future are the so-called absolute laws of nature, which we have mastered over hundreds of years of life on this planet through rational processes. These robust laws, ruthless in their accuracy, can predict the exact date of future eclipses or even the impending approach of an asteroid or comet. Yet, some researchers are now concerned about the immutability of such laws. Are these laws unchanging and absolute, or are they evolving with time like the stars and planets?

We like to see our world as deterministic, because it has innate beauty and order. It makes sense to think that the cause precedes the effect. Even if our universe is a random accident, we still want to believe that it must have been caused by the deterministic laws that govern it. Stephen Hawking once wrote, "I have noticed even people who claim everything is predestined, and that we can do nothing to change it, look before they cross the road."

But, what if these laws of physics and constants, considered as sacred, vary over a long period of time? Or, even worse, what if they are different in distant parts of the universe. Then, one has to say the random universe is governed by apparent laws rather than absolute ones. Scientists depend heavily on these laws to interpret observations and make predictions. And, if these laws are not absolute, we would create a different history for our universe based on those findings. Truly, "the observation creates history."

Even in the technology-driven twenty-first century, horoscopes serve as a daily menu to readers of our newspapers. The basic premise of this old tradition, broadly speaking, is that the Sun and other celestial bodies have an impact on our daily lives. The surprising ability of this ancient notion of astrology to survive the scientific and technological evolution is a puzzle. Interestingly, astrology defies any rational or critical thinking, yet it survives.

The suffix-*ology* denotes a branch of knowledge or science. In that sense astrology may seem to fit easily into the family of sciences, like biology, geology, psychology, etc. And this was true for a long time, in fact, for most of human history. However, with the advent of modern physics, the burden and onus of proof for any celestial phenomena have had to be involved with the laws of nature or physics. And, it looks like our Earthly life is too complicated to be dictated by celestial spheres.

Although the name astrology still remains in place, this particular branch of knowledge either lost all its characteristics or it never possessed those that could qualify it as a science. Astrology is somewhat like a fossil that lays buried in the vast landscape of modern science. Adding flesh to this old skeleton is not enough to equip it to stand the vigorous and rational standards of science. It was founded on a belief system rather than on inquiry-based observations. Its core base is anecdotal evidence – stories that have been shared by generations.

Astrology has many of the accessories that resemble real science, like complicated diagrams and a dedicated vocabulary, but astrologers do not follow any logical method. For the most part, they lack the audacity to challenge their own findings. They are alarmed at the possibility that any unfavorable outcomes will shake the foundation of their deeply held convictions.

This is in no way to suggest that there is no correlation between terrestrial and celestial events. We can predict the timing of eclipses and tides on Earth for a long time to come, a clear indication of the indisputable relationship between the celestial and terrestrial. But, here is the major difference: such conclusions, in the scientific realm, are based on testable and consistent theories and laws that can be confirmed by observations and repeated at will. In principle, this connection is the basic foundation of astrology and its modern version. However, astrology somehow is in short supply of any acceptable or consistent methodologies that can be employed to explore the unknown. The arsenal of astrology contains outdated tools such as vague explanations and biased presumptions.

Astrology never came out of the ancient pit where all the branches of science once resided. Although astrology, similar to any science, seeks to explain the natural world, astrologers don't usually attempt to critically evaluate their findings. In the first place, they have immense (irrational) belief in the limited and incomplete foundation on which the basic conjunctures are built.

Science is also plagued by occasional frauds, and many of its findings are based on primitive assumptions. It's not the final word on every matter, either. However, it is the best tool we have, though not a perfect one, to seek the unknown.

In science, testable ideas and critical thinking are not only the part of its core, but the rewriting and reinventing of old ideas and new level of understanding are essential units of its growth. Even well-established theories are susceptible to strong evidence, and we may have to give them up even if they are dear to us.

Science doesn't have irrefutable definitions or final answers for many observed phenomena. That's also the strength of science. But, this is not a reason to accommodate anything like astrology under its umbrella. Scientific pursuits have common features that are not quite observable and that have never been a major part of astrology.

Even if newspapers free up the square inches devoted to horoscopes, they might still have a place in the minds of the people. It is one of those little ironies of life, and it does not vanish so quickly.

As we continue our scientific voyage into the unchartered terrains of the cosmos, it also instills a sense of humility in us. The universe came from nothing and it must end in a similar state, though we are now aware that nothingness is not what we thought earlier!!

Why should we weigh scientific reasoning more than any other thoughts, even though science has and accepts its limitations? Scientific pursuit and wisdom take us forward, not back to the dark ages.

It is surprising to know that many segments of society remain ignorant or pretend to be ignorant in scientific matters. Our civilization, in order to march forward, must spend more energy and effort on scientific enterprises than trivial matters. Many scientists including Einstein had deep interest in matters that related to society.

FROM THE DUST WE CAME AND TO THE DUST WE SHALL RETURN

Astronomers have for decades suspected that exploding stars, known as supernovae, are the dust-making machines of the early universe. This assumption has been vindicated by the most recent observations from the Herschel Space Observatory.

The supernova explosion dubbed as SN 1987A, as the name indicates, was seen on Earth in 1987 with the naked eye. It's believed to have occurred about 170,000 light-years away in a neighboring galaxy and was one of the brightest cosmic explosions in recorded history.

The Herschel Space Observatory, operated by the European Space Agency, revealed the enormous amount of cosmic dust spewed by this explosion. The leftover dust from this blast, until now, has been largely buried under the cosmic carpet. However, the latest findings have surprised the astronomers bigtime. The vast quantity of dust, generated in this event, is equivalent to about 200,000 Earth masses.

This colossal amount of dust from just one supernova explosion elucidates the abundance of this cosmic stuff in galaxies. In fact, it demystifies the puzzle that surrounds the origin of cosmic dust in galaxies.

It's been known for a long time that we are truly made of star stuff. The carbon in our body cells, the calcium in our bones, and the iron in our blood – all came from exploding stars. The early universe was filled with the cosmic dust from such gigantic explosions. From that dust came the next generation of stars, planets and all other creatures including us.

The cosmic dust includes carbon, nitrogen, oxygen, calcium, iron and various other elements. As hard as it might be to believe, all atoms in our body, except hydrogen, originated billions of years ago in stars. We are the aftermath of these stellar explosions. These atoms that make up our body were engineered inside stars about 13 billion years ago. And,

Figure 12.6. This artist's impression shows the disc of gas and cosmic dust around a brown dwarf, a star-like object, but one too small to shine brightly, like a star. The rocky planets like earth may be even more common in the universe than expected (Image credit: NASA).

the hydrogen in our body was created even before this – in the Big Bang about 13.7 billion years ago. During the grand cycle of birth and death, these atoms have been recycled many times, and even they were part of rocks, plants and animals in the past before becoming part of our own body.

When we depart, the very same atoms that make us now will continue their never-ending saga to breed new avatars of stars, planets, and people. They will be churned by the forces of nature like clay in the potter's hand. Interestingly, from the dust we came and to the dust we shall return!

REFERENCES

About Chandrayaan-1. (2013). About Chandrayaan-1. Available at http://www.isro.org/Chandrayaan/htmls/about_chandrayaan.htm. Accessed 30 June 2013.

Allison, W. (2009). *Radiation and reason: Impact of science on a culture of fear.* York: York Publishing Services. http://www.radiationandreason.com

Bojowald, M. (2006). Universe scenarios from loop quantum cosmology. *Annalen der Physik, 15*(4–5), 326–341.

Dimmitt, C. (1978). *Classical Hindu mythology: A reader in the Sanskrit puranas.* Philadelphia: Temple University Press.

Einstein, A. (1995). *Relativity: The special and the general theory.* Reprint Edition. New York: Broadway Books.

Frost, R. (1920). "Fire and Ice" a group of poems. *Harper's Magazine.* December.

Greene, B. (1999). *The Elegant Universe: Superstrings, Hidden Dimensions, and the Quest for the Ultimate Theory,* New York, W. W. Norton.

Hey, T. (2003). *The new quantum universe* (Revised and Updatedth ed.). New York: Cambridge University Press.

Källne, J., et al. (2012). Fusion for neutrons and subcritical nuclear fission: Proceedings of the international conference (AIP conference proceedings/plasma physics). 2013 Edition. Melville: American Institute of Physics.

Laughlin, R. B. (2006). *A different universe: Reinventing physics from the bottom down.* New York: Basic Books.

Lord George, G. B. (2002). *Selected poetry of Lord Byron (Modern library classics).* (New Ed.). Modern Library.

Newton, I. (2000). Notes on early church history and the moral superiority of the 'barbarians' to the Romans (Normalized version). Welcome|Newton project. http://www.newtonproject.sussex.ac.uk/view/texts/normalized/THEM00076. Accessed 4 July 2012.

Nuclear Energy: Fission and Fusion – Harvard – Belfer Center for Science and International Affairs. (2013). Nuclear energy: Fission and fusion – Harvard – Belfer Center for Science and International Affairs. http://belfercenter.hks.harvard.edu/publication/2246/nuclear_energy.html. Accessed 02 July 2012.

Nuclear Power Plants|RadTown USA|US EPA. (2013). Nuclear power plants|RadTown USA|US EPA. http://www.epa.gov/radtown/nuclear-plant.html. Accessed 02 July 2013.

Pieters, C. M., et al. (2009). *Science, 326,* 568–572.

Robert, B. L. (2008). *The crime of reason: And the closing of the scientific mind.* New York: Basic Books.

The Hebrew University of Jerusalem. Department of Media Relations. (2009). The Hebrew University of Jerusalem. Department of Media Relations. http://www.huji.ac.il/cgi-bin/dovrut/dovrut_search_eng.pl?mesge118215035732688760. Accessed 12 June 2009.

Wagemans, C. (1991). *The nuclear fission process* (1st ed.). Boca Raton: CRC Press.

Weinberg, S. (1994). *Dreams of a final theory: The scientist's search for the ultimate laws of nature,* (3rd Ed.). New York: Vintage.

World Nuclear Association. (2013a). World Nuclear Association. http://www.world-nuclear.org. Accessed 30 June 2013.

World Nuclear Association. (2013b). World Nuclear Association. http://world-nuclear.org/. Accessed 02 March 2011.

Final Words

Will we find a theory of everything? Though efforts are under way to know the universe, we are quite far from achieving the ultimate truth.

My 9-year-old son Nathan believes his brother Noah has become a star in the sky after his demise. In my view, this is a simple and powerful way of connecting to the cosmos, though children might have learned such metaphorical stories from fairy tales or stories told by elders.

We are linked to the stars in many ways. All the chemical elements that make us up came from the stars, and when we complete the terrestrial cycle, the fundamental particles must become part of something else. The cycles of creation and destruction will go on. Given enough time, our species must be able to discover many things that we can't even imagine now, assuming we don't destroy ourselves and we have the knowhow to escape the wrath of the destructive forces that have been unleashed on our planet.

In his 1956 short story, "The Last Question," Isaac Asimov takes the idea of future to a logical extreme. The story narrates that our descendants become advanced "energy beings" in the far future. They are able to fabricate a hyper-dimensional super-computer that essentially can reboot the unwinding universe. They asked the computer: "How can the net amount of entropy of the universe be massively decreased?" In other words, how can the workings of the universe be reversed? Each time, the response from the computer was "There was insufficient data for meaningful answer." Eventually, humanity is gone, and the universe is dead. The descendants of the computer that has been engaged in figuring out the answer all this time finally came up with an answer, but no one was around to accept it. So it decided to perform the reversal of the universe, and there is a second Big Bang when the computer asserts: "Let there be light!" And there was light!

Our intelligence may be augmented by the machines, they may even transcend us, but the fundamental questions remain the same.

In the last decade, the much-celebrated string theory has come to the fore. The string theory replaces zero-dimensional particles with one-dimensional string as the building blocks of matter. The vibrations of these extremely small strings manifest as matter and energy to create everything in the universe. String theory offers a ten-dimensional space-time instead of the familiar four-dimensional space-time. These extra dimensions are hard to conceive, let alone test for in a laboratory setup. The inability to provide any

S. Mathew, *Essays on the Frontiers of Modern Astrophysics and Cosmology*, Springer Praxis Books, 199
DOI 10.1007/978-3-319-01887-4, © Springer International Publishing Switzerland 2014

quantitative experimental evidence is the main criticism against string theory. Some detractors even call it a theory of philosophy rather than physics.

In spite of all these limitations, string theory enjoyed wide participation from the greatest minds of our generation. Their goal was to unify the existing theories and fulfill Einstein's dream of the "theory of everything."

The emergence of different string theories in the last decades was misleading for scientists who have been looking for a cohesive, all-encompassing picture of the cosmos. Mathematically, all different types of string theories are equally acceptable, yet they differ in some features. This has led Edward Witten of the Institute of Advanced Study, known to many as the most brilliant physicist of our generation, to announce the M-theory in 1995. It was an extension of different string theories and sparked the second superstring revolution. The M-theory introduced membranes as the fundamental entities of existence, and it functions in an 11-dimensional space-time.

The existing experimental facilities may not be able to test string theory or M-theory in the near future. Some argue that such a scenario would qualify it to be considered as a mathematical framework rather than a physical theory. On the other hand, the large number of possible solutions offered by these theories has the potential to explain any phenomena that is beyond our current theoretical structure. Many scientists, especially the proponents of string theory, expect that the high-energy experiments scheduled at the Large Hadron Collider (LHC) may verify at least some aspects of the M-theory, possibly extra dimensions. Surprisingly, the existence of parallel universes is an undemanding result of the string theory, and to deny that, one needs to further complicate the theory.

We may not be able to answer the last question. Yet we are here for a short period of time to fine tune that final answer and must share our knowledge with our descendants, and they will continue their voyages to territories that are unknown to us. We are destined to do that before we depart to become star stuff. Science, as nothing else can do, teaches us humility while we seek the truth.

Index

S. Mathew, *Essays on the Frontiers of Modern Astrophysics and Cosmology*, Springer Praxis Books, 201
DOI 10.1007/978-3-319-01887-4, © Springer International Publishing Switzerland 2014